NANOTECHNOLOGY SCIENCE AND TECHNOLOGY

# NANOTECHNOLOGY IN DRUG DELIVERY

## STRATEGIES, TECHNOLOGIES AND APPLICATIONS

# NANOTECHNOLOGY SCIENCE AND TECHNOLOGY

Additional books in this series can be found on Nova's website under the Series tab.

Additional e-books in this series can be found on Nova's website under the e-book tab.

NANOTECHNOLOGY SCIENCE AND TECHNOLOGY

# NANOTECHNOLOGY IN DRUG DELIVERY

## STRATEGIES, TECHNOLOGIES AND APPLICATIONS

PROF. A. K. HAGHI
UNIVERSITY OF GUILAN, RASHT, IRAN
EDITOR-IN-CHIEF, POLYMERS RESEARCH JOURNAL (NOVA)

PROF. E. A. CASTRO
RESEARCH INSTITUTE OF THEORETICAL AND APPLIED PHYSICAL-CHEMISTRY (INIFTA), ARGENTINA

nova publishers
*New York*

**Library of Congress Cataloging-in-Publication Data**

Polystyrene : properties, performance, and applications / editors, James E. Gray.
p. cm.
Includes index.
ISBN: 978-1-62948-425-9 (hardcover)
1. Polystyrene. I. Gray, James E. (James Ehren), 1960-
TP1180.S7P75 2010
668.4'233--dc22
2010047082

*Published by Nova Science Publishers, Inc. † New York*

# CONTENTS

# PREFACE

Technical and technological development demands the creation of new materials, which are stronger, more reliable and more durable, i.e. materials with new properties. Up-to-date projects in creation of new materials go along the way of nanotechnology.

Nanotechnology can be referred to as a qualitatively new round in human progress. This is a wide enough concept, which can concern any area: information technologies, medicine, military equipment, robotics and so on. We narrow the concept of nanotechnology and consider it with the reference to polymeric materials as well as composites on their basis.

The prefix nano- means that in the context of these concepts the technologies based on the materials, elements of constructions and objects are considered, whose sizes make $10^{-9}$ meters. As fantastically as it may sound but the science reached the nanolevel long ago. Unfortunately, everything connected with such developments and technologies for now is impossible to apply to mass production because of their low productivity and high cost.

That means nanotechnology and nanomaterials are only accessible in research laboratories for now, but it is only a matter of time. What sort of benefits and advantages will the manufacturers have after the implementation of nanotechnologies in their manufactures and starting to use nanomaterials? Nanoparticles of any material have absolutely different properties rather than micro-or macroparticles. This results from the fact that alongside with the reduction of particles' sizes of the materials to nanometric sizes, physical properties of a substance change too.

Such changes in properties of substances are explained by the quantitative change of surface and volume atoms' ratio of individual particles, i.e. by the high-surface area. Insertion of such nanoparticles in a polymeric matrix while using the apparently old and known materials gives a chance of receiving the qualitatively and quantitatively new possibilities in their use.

The desired event of fast implementation of nanomaterials in mass production depends on the efficiency of cooperation between the scientists and the manufacturers in many respects. Today's high technology problems of applied character are successfully solved in close consolidation of scientific and business worlds.

**A. K. Haghi,** University of Guilan, Rasht, Iran
**E. A. Castro,** Research Institute of Theoretical and Applied Physical-Chemistry (INIFTA), Argentina

*Chapter 1*

# ORAL DELIVERY SYSTEMS

## COLON TARGETED DRUG DELIVERY SYSTEMS

The oral route for drug administration is the most convenient and important method in administering drugs for a systemic effect. Nearly 50% of the drug delivery systems available on the market today are oral D.D.S. These methods provide more advantages due to patient acceptance and ease of administering the necessary drugs. There are various methods or techniques through which colon drug targeting can be achieved. For example, drug targeting methods pertain to the formation of prodrug, coating with pH-sensitive polymers, coating with biodegradable polymers, designing formulations using polysaccharides, timed released systems, pressure-controlled drug delivery systems, and osmotic pressure controlled systems. Coating of the drugs with pH-sensitive polymers provides a simple approach for colon-specific drug delivery. However, reaching the proximal part of the colon via rectal administration is difficult. Rectal administration can also be uncomfortable for the patient, and compliance may be less than optimal.

There are several ways in which drugs can be targeted for the colon when they are given by mouth. In time-dependent formulations, the drug concerned is released during the period of gastrointestinal transit time. Release from formulations that contain pH-dependent polymers takes place on the basis that pH is higher in the terminal ileum and colon than in the upper parts of the gastrointestinal tract. Prodrugs and dosage forms that account for when the drug release is triggered by the action of colonic bacterial enzymes have therefore been devised.

The colon is in the lower part of the gastrointestinal tract and runs from the ileocecal junction valve to the anus. This area also includes the proximal part (ascending colon), transverse colon, descending colon, sigmoid colon, rectum, and anus (Figure 1).

In the colon, the surface area of the small intestine is effective since this is where absorption takes place due to the presence of villi, microvilli, and the function process has a long residence time. The colon is a cylindrical tube which is lined by a moist, soft pink lining called mucosa that is about 2-3 inches in diameter. The colon and rectum have an anatomic blood supply. Lymph nodes are also present with the blood vessels.

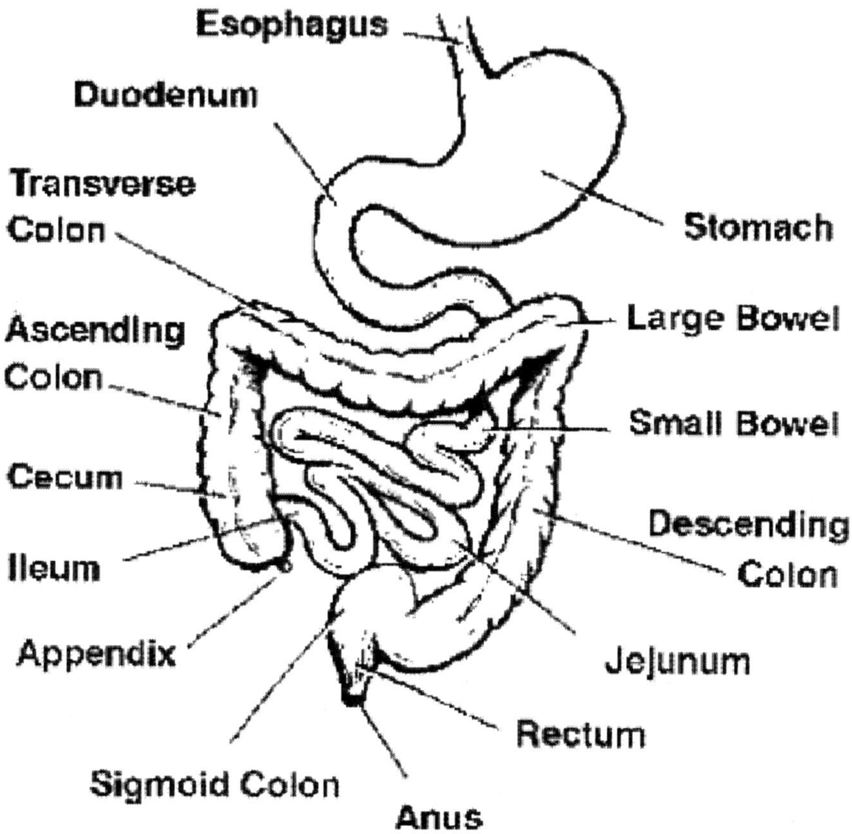


Figure 1. This diagram illustrates the various regions in the gastrointestinal tract.

Activity in the colon can be divided into segmenting and propulsive movements. Segmenting movements are caused by a circular muscle that has the appearance of a sac-like haustra, which result in mixing the luminal contents. Significant propulsive activity, associated with defecation and affected by a longitudinal muscle is less common and occurs at an average of three or four times daily.

The slow movement of material through the colon allows a large microbial population to grow there. Over 400 distinct bacterial species have been found. Most of these isolated bacteria are anaerobic in nature. A small number of fungi are also present. The rate of microbial growth is greatest in the proximal areas because of the high concentration of energy.

The principal source of nutrition for the colonic microorganisms is carbohydrates arriving in intestinal chime. The carbohydrates are degraded by the action of the polysaccharidase and glycosidase enzymes and the ultimate products of this fermentation are short chain fatty acids. The carbohydrate fermentation predominates in the colon which results in a relatively low pH. In the distal regions, there is little carbohydrate fermentation, resulting in a higher pH. The bacteria within the colon are predominantly anaerobic and there is a low redox potential.

## ABSORPTION OF DRUGS FROM THE COLON

Drugs are absorbed passively by paracellular or transcellular routes. Transcellular absorption involves the passage of drugs through cells and this is the route most lipophilic drugs take; whereas paracellular absorption involves the transport of the drug through the tight junctions between cells and is the route that most hydrophilic drugs take. The poor paracellular absorption for many drugs in the colon is due to the fact that epithelial cell junctions are very tight.

### Angiodysplasia

Tortuous dilution of sub-mucosal and mucosal blood vessels is seen most often in the cecum or right colon, usually after the age of 60. These vessels are prone to rupture and bleed into the lumen. These lesions account for 20% of significant lower intestinal bleeding. Angiodysplasia is a small vascular malformation of the gut. It is a common cause of otherwise unexplained gastrointestinal bleeding and anemia. Lesions are often multiple, and frequently involve the cecum or ascending colon, although they can occur at other places. Treatment may be with endoscopic interventions, medication, or occasionally surgery. Diagnosis of angiodysplasia is often accomplished with endoscopy, either colonoscopy or esophagogastroduodenoscopy (EGD).

### Inflammatory Bowel Disease

Crohn's disease may affect any portion of the gastrointestinal tract from the esophagus to the anus but most often involves the ileum. There are multiple factors that can cause inflammatory bowel disease and symptoms can come from various kinds of inflammatory responses, i.e., abnormal local immune response against the normal flora of the gut, genetic factors such as multiple genetic factors, candidate genes, and chromosome location. Bowel disease can also be caused by other infectious agents like Escherichia coli, Measles virus, Cytomegalovirus, or dietary factors such as saturated fats, milk products, allergic foods, etc. Crohn's disease and ulceration colitis are chronic relapsing inflammation disorders with unknown origins that are collectively known to be referred to as an idiopathic inflammatory bowel disease (IBD). The main drugs used in the treatment of ulcerative colitis and Crohn's disease are the amino salicylates and corticosteroids. These diseases along with other inflammatory bowel diseases have been linked to an increased risk for colorectal cancer.

*Ulcerative colitis*: Ulcerative colitis occurs only in the large intestine. Ulcers form in the inner lining of the intestine, or mucosa, in the colon or rectum, often resulting in diarrhea, blood, and pus. The inflammation is usually very rigorous in the sigmoid and rectum, and usually reduces in the colon.

*Crohn's disease*: Crohn's disease, also called regional enteritis, is a chronic inflammation of the intestines which is usually confined to the terminal portion of the small intestine, or the ileum.

## Colorectal Cancer

Large bowel cancer includes cancerous growths in the colon, rectum, and appendix. Statistics have shown that 98% of all cancers in the large intestine are adenocarcinomas. Several studies suggested that the use of aspirin and other NSAIDs have a protective effect against colon cancer. Colorectal cancers arise from adenomatous polyps in the colon. These mushroom-shaped growths are usually benign, but some develop into cancer over time. Localized colon cancer is usually diagnosed through colonoscopy. Invasive cancers that are confined within the wall of the colon (TNM stages I and II) are curable with surgery. If untreated, they spread to regional lymph nodes (stage III), where up to 73% are curable by surgery and chemotherapy. Cancer that metastasizes to distant sites (stage IV) is usually not curable, although chemotherapy can extend survival. And, in rare cases, both surgery and chemotherapy together, have been shown to cure some patients. Radiation is used with rectal cancer.

## Drugs Used in Colon Cancer

5-fluorouracil

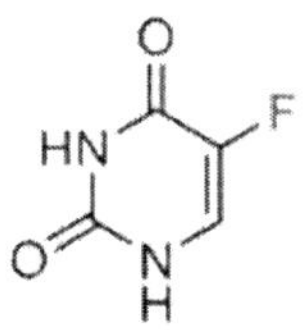


5-Fluorouracil

9-aminocamptothecin

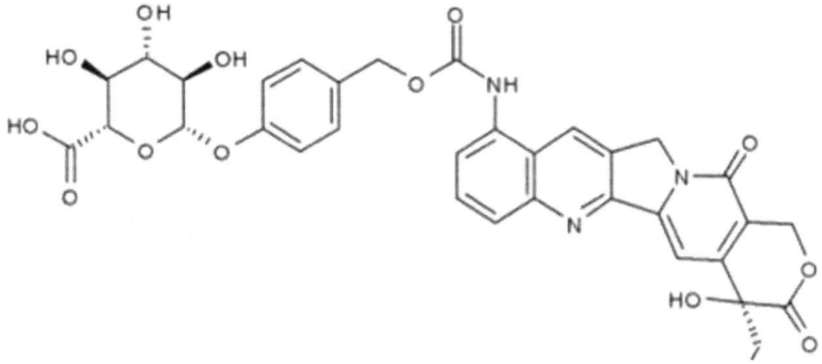

Capecitabine

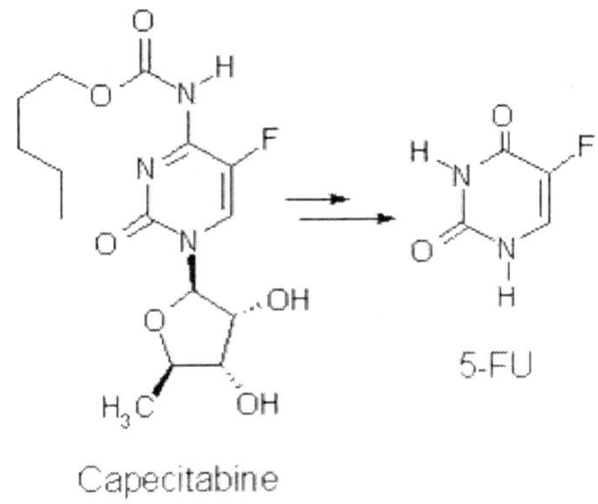


Cetuximab

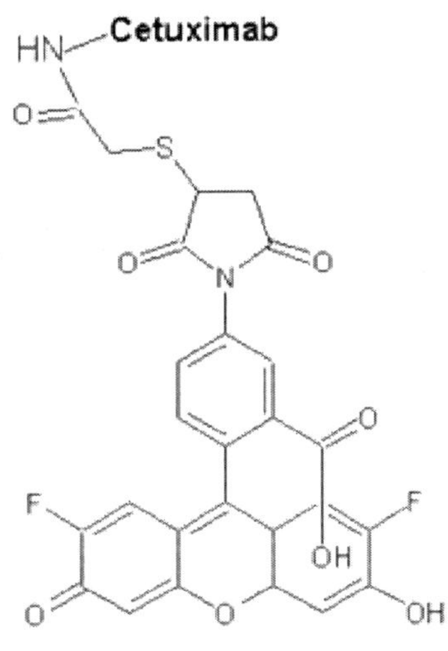


Trinotecan

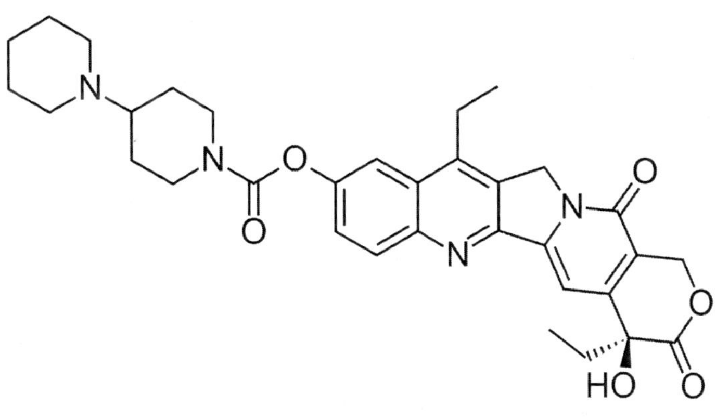

Levamisole hydrochloride

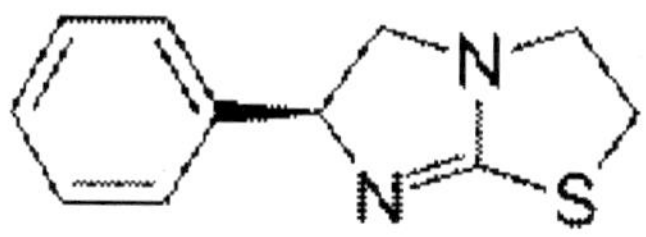


Oxaliplatin

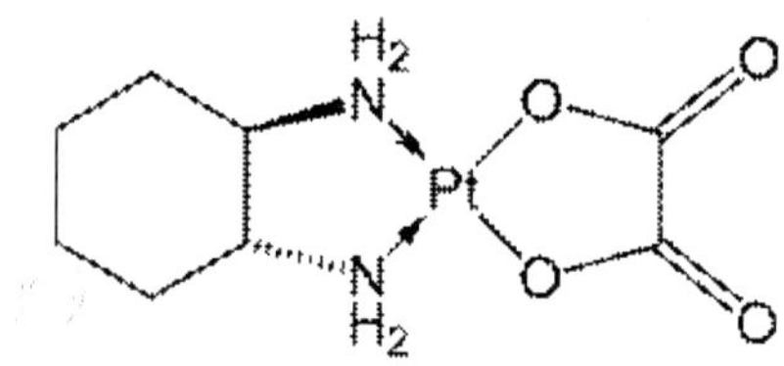


Trimetrexate

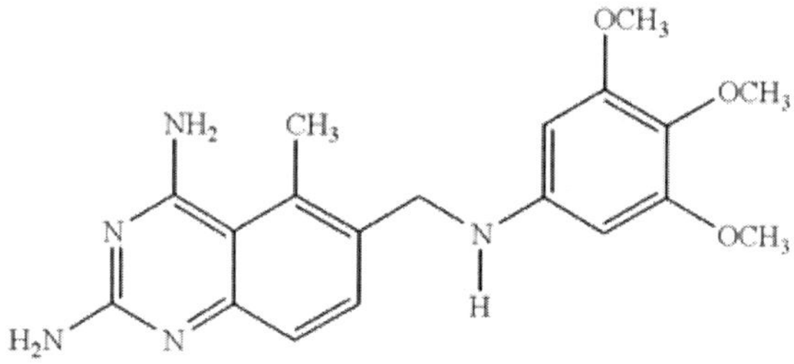


UFT (ftorafur and uracil)

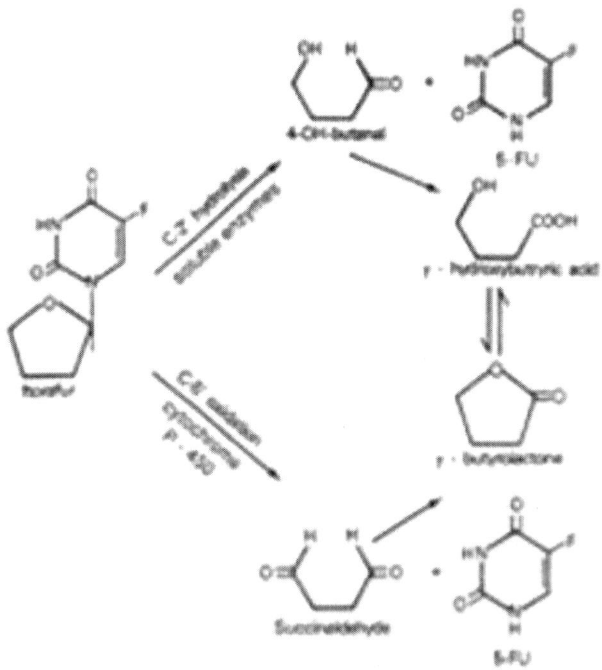

Bevacizumab

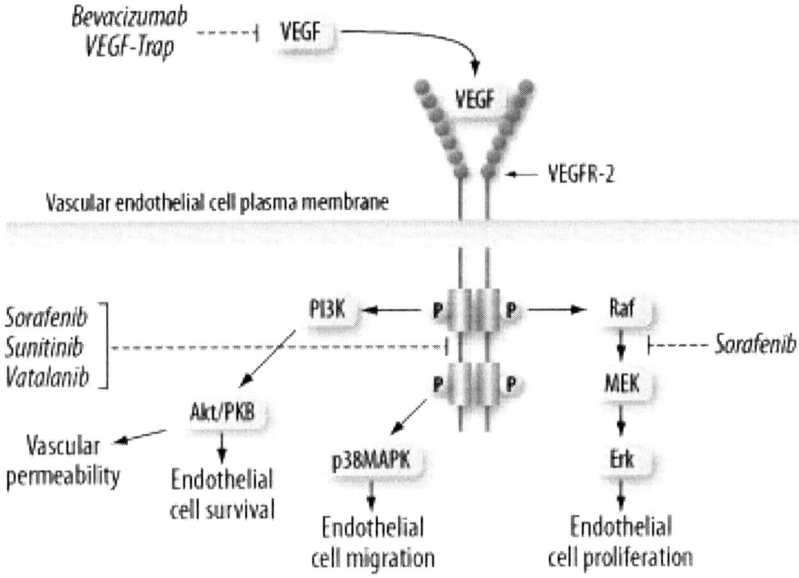


Cisplatin

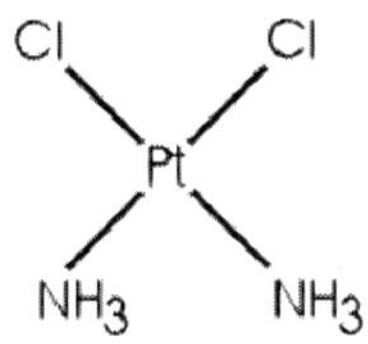


Cisplatin

## Constipation

Constipation (also known as costiveness, dyschezia, and dyssynergic defecation) refers to bowel movements that are infrequent and hard to pass. Constipation is a common cause of painful defecation. Severe constipation includes obstipation and fecal impaction. Treatments include changes in dietary habits, laxatives, enemas, biofeedback, and surgery. Because constipation is a symptom, and not a disease; effective treatment of constipation may require investigating the symptoms of its underlying cause.

## Diarrhea

Diarrhea is the condition of having three or more loose or liquid bowel movements per day. The loss of fluids through diarrhea can cause dehydration and electrolyte imbalances. Inflammatory diarrhea occurs when there is damage to the mucosal lining or brush border, which leads to a passive loss of protein-rich fluids, and a decreased ability to absorb these lost fluids. Features of all three of the other types of diarrhea can be found in this type of diarrhea. It can be caused by bacterial infections, viral infections, parasitic infections, or autoimmune problems such as inflammatory bowel diseases. It can also be caused by tuberculosis, colon cancer, or enteritis.

## Diverticulitis and Diverticulosis

A diverticulum is a blind pouch that communicates with the lumen in the gut. Congenital diverticula can involve all three layers of the bowel wall and is distinctly uncommon. Acquired diverticula may occur anywhere in the alimentary tract, but by far the most common location is in the colon. Diverticulitis can occur when the diverticula becomes inflamed. An initial episode of acute diverticulitis is usually treated with bowel rest (*i.e.*, nothing by mouth), IV fluid resuscitation, and broad-spectrum antibiotics which cover anaerobic bacteria and gram-negative rods. However, recurring acute attacks or complications, such as peritonitis, abscess, or fistula may require surgery, either immediately or on an elective basis.

## Hirschsprung's Disease (Aganglionosis)

Hirschsprung's disease results when, during development, the migration of neutral crest-derived cells along the alimentary tract arrests at some point before reaching the anus. The critical lesion in Hirschsprung's disease is the lack of ganglion cells, and ganglia, in the muscle wall and sub-mucosa of the affected segment. Hirschsprung's disease is a rare congenital (present at birth) abnormality that results in obstruction because the intestine does not work normally. It is most often found in males. It is commonly found in Down syndrome children. It can be life-threatening or a chronic disorder. In a newborn, the chief signs and symptoms are failure to pass a meconium stool within 24-48 hours after birth, reluctance to eat, bile-stained (green) vomiting, and abdominal distension. During infancy the child has difficulty gaining weight, constipation, abdominal distension, episodes of diarrhea, and vomiting. Explosive watery diarrhea, fever and exhaustion are signs of enterocolitis (inflammation of the colon) and are considered serious and life-threatening. If these symptoms occur, notify your child's doctor immediately. In older children, symptoms become chronic and include constipation, passage of ribbon-like, foul-smelling stools, abdominal distension and visible peristalsis (wave-like movement of the intestines). An older child is usually poorly nourished and anemic.

## Ileus

Ileus is commonly defined as an intestinal obstruction. Ileus is a disruption of the normal propulsive gastrointestinal motor activity due to non-mechanical causes. In contrast, motility disorders that result from structural abnormalities are termed as a mechanical bowel obstruction. Ileus has three types: Postoperative Ileus, Paralytic Ileus, and Acute colonic pseudo-obstruction.

## Intussusception

An intussusception is a medical condition in which a part of the intestine has invaginated into another section of intestine, similar to the way in which the parts of a collapsible telescope slide into one another. The telescoped segment is called the intussusceptum and lower receiving segment is called the intussuscipiens. This region of the intestine can often result in an obstruction. The part that prolapses into the other portion of the intestine is called the intussusceptum, and the part that receives it is called the intussuscipiens. The condition is most common in infants and children. The condition is not usually immediately life-threatening. The intussusception can be treated with either barium or water-soluble contrast enema or an air-contrast enema, which both confirms the diagnosis of intussusceptions and in most cases successfully reduces it. The success rate is over 80%. However, approximately 5-10% of these recur within 24 hours.

## Irritable Bowel Syndrome

Irritable bowel syndrome (IBS) or spastic colon is a diagnosis of exclusion. It is a functional bowel disorder characterized by chronic abdominal pain, discomfort, bloating, and alteration of bowel habits in the absence of any detectable organic cause. IBS may begin after an infection, or a stressful life event.

Although there is no cure for IBS, there are treatments that attempt to relieve symptoms, including dietary adjustments, medication, and psychological interventions. Patient education and a good doctor-patient relationship are also important. Several conditions may be present with IBS, including celiac disease, Fructose malabsorption, mild infections, or parasitic infections like giardiasis, several inflammatory bowel diseases, functional chronic constipation, and chronic functional abdominal pain. In IBS, routine clinical tests yield no abnormalities, though the bowels may be more sensitive to certain stimuli, such as the balloon insufflation testing. The exact cause of IBS is unknown. The most common theory is that IBS is a disorder that deals with the interaction between the brain and the gastrointestinal tract, although there may also be abnormalities in the gut flora or the immune system.

## Pseudomembranous Colitis

Pseudomembranous colitis, also known as antibiotic-associated diarrhea (AAD), is an infection of the colon. It is often, but not always, caused by the bacterium Clostridium difficile. The illness is characterized by offensive-smelling diarrhea, fever, and abdominal pain. In severe cases, life-threatening complications can develop, such as toxic mega-colon.

## Hemorrhoids

Hemorrhoids or piles are the varicosities of the hemorrhoidal veins. Lesions are often found in the elderly and pregnant women. Hemorrhoids commonly result from an increase in venous pressure. The possible causes include portal hypertension, chronic constipation, and straining to eliminate stool, cardiac failure, venous stasis of pregnancy, hereditary predisposition, or with tumors in the rectum.

# NEW TARGETED DRUG DELIVERY SYSTEMS

Colorectal cancer is the second leading cause of cancer related deaths for both men and women. Each year there are about one million new cases of colorectal cancer. Its incidence has increased over the last 25 years. It has been estimated that about 135,000 new cases of colon cancer occurred in the United States during 2001; and 141,405 new cases, and 53,000 deaths from colorectal cancer in 2005. In Saudi Arabia, cases of colorectal cancers constituted 5.5% of the total cancer cases reported. In 2008, colorectal cancer was the number one cancer in males according to the National Cancer Registry at King Faisal Specialist Hospital and Research Center, and the second in females after breast cancer.

Colon cancer is a disease in which malignant cells form in the tissues of the colon. These abnormal cells create tumors and are characterized by unregulated replication and they may spread to other sites. Mortality rate associated with this type of cancer is high in industrialized areas and lower in rural areas. The early detection and diagnosis and the use of less toxic systems would tremendously improve the efficacy of therapy.

## Drug Release Based on Variation of pH

The stomach pH ranges between 1 and 2 during fasting but will increase after eating. The pH is about 6.5 in the proximal small intestine and about 7.5 in the distal small intestine. From the ileum to the colon, pH declines significantly. The pH is about 6.4 in the cecum. However, pH values as low as 5.7 have been measured in the ascending colon in healthy volunteers. The pH in the transverse colon is 6.6, in the descending colon 7.0. Use of pH-dependent polymers is based on the aspect of these pH level differences. The polymers described as pH-dependent in colon specific drug delivery are insoluble at low pH levels but become increasingly soluble as the pH level rises. There are various problems with this approach, however. The pH in the gastrointestinal tract varies between and within individuals. It is affected by diet and disease, for example. During acute stages of inflammatory bowel disease, colonic pH has been found to be significantly lower than normal. In ulcerative colitis pH values between 2.3 and 4.7 have been measured in the proximal parts of the colon. Although a pH-dependent polymer can protect a formulation in the stomach and proximal small intestine, it may start to dissolve even in the lower small intestine, and the site-specificity of formulations can be poor. Contrariwise, failure of enteric-coated dosage, especially with single-unit dosages was because of the lack of disintegration reported. The decline in pH from the end of the small intestine to the colon can also result in problems. Lengthy lag times at the ileocecal junction or rapid transit through the ascending colon can also result in poor site-specificity of enteric-coated single-unit formulations. Eudragit™ products are pH-dependent methacrylic acid polymers containing carboxyl groups. The number of esterified carboxyl groups affects the pH level at which dissolution takes place.

Eudragit™ S is soluble above pH 7 and Eudragit™ L above pH 6. Eudragit™ S coatings protect well against drug liberation in the upper parts of the gastrointestinal tract and have been used in preparing colon-specific formulations. When sites of disintegration by Eudragit™ S-coated single-unit tablets were investigated using a gamma camera; they were found to lie between the ileum and splenic flexure. Site specificity of Eudragit S formulations, in both single- and multiple-units were usually poor.

Eudragit™ S coatings have been used to target the anti-inflammatory drug 5-aminosalicylic acid (5-ASA) in single-unit formulations on the large intestine. Eudragit™ L coatings have been used in single-unit tablets to target 5-ASA on the colon in patients with ulcerative colitis or Crohn's disease. The polypeptide hormone vasopressin and insulin have been administered to rats orally in Eudragit™ S-coated single-unit capsules. Eudragit™ S-coated insulin capsules have also been administered orally to hyperglycemic beagle dogs. In the latter study, it was concluded that plasma glucose levels were lowered gradually and reproducibly but that delivery by means of the oral route was not bioequivalent to delivery by means of parenteral route (SC). Eudragit™ S has been used in combination with another methacrylic acid copolymer, Eudragit™ L100-55, in colon-targeted systems to regulate drug delivery. Dissolution studies showed that drug release profiles from enteric-coated single-unit tablets could be altered *in vitro* by changing the ratios of the polymers, in the pH range 5.5 to 7.0. Hydroxypropylmethylcellulose acetate succinate (HPMCAS) has been included in outer layers of single-unit press-coated tablets with the view of preventing drug release in the stomach and small intestine. *In vitro* dissolution studies suggested that such tablets could be useful as colon-specific formulations. No *in vivo* studies were undertaken.

## Drug Release Based on Gastrointestinal Transit Time

The time in transit through the small intestine is independent of formulation. It has been found that both large single-unit formulations and small multiple-unit formulations take three to four hours to pass through the small intestine (Davis et al., 1986; Parker et al., 1988; Wilson et al., 1989a; and Adkin et al., 1993). Transit time through the small intestine is unaffected by particle size or density, or by the composition of meals. Because the time taken by formulations to leave the stomach varies greatly since the arrival time for the formulation in the colon cannot be accurately predicted. However, the effects of this variation in gastric residence time can be *minimized* by using systems that are protected in the stomach. Drug release can be targeted on the colon by means of formulations that release the drug they contain at a certain time after gastric emptying. Such formulations pass through the stomach and small intestine and the drug is then released at the end of the small intestine or in the beginning of the colon (Kinget et al., 1998). Accordingly, formulations that depend for a drug release on the time of transit through the small intestine, also usually depend on the drug release to be triggered on changes in pH within the gastrointestinal tract. Transit times through the colon that are faster than normal have been observed in patients with irritable bowel syndrome, diarrhea, and ulcerative colitis. Systems that depend on gastrointestinal transit time for drug release are therefore not ideal for drug delivery in the colon for treatment of a colon-related disease.

Combinations of hydrophilic (hydroxypropylmethylcellulose, HPMC) and hydrophobic polymers have been used as coatings for tablets that release the drug from a core after a lag time (Pozzi et al., 1994). When the *in vivo* behavior of such tablets was studied scintigraphically it was found that disintegration occurred in the proximal colon after about 5.5 hours (range 5 to 6.5 hours). Lag time could be adjusted by changing the thickness of the polymer layer. HPMC and hydroxypropylcellulose (HPC) have been used as swellable polymers in delayed release formulations. In such formulations enteric polymers can also be used as coatings to protect the formulation in the stomach. Using gammascintigraphy, Sangalli et al., investigated the *in vivo* behavior of tablets with a drug-containing core coated with hydrophilic HPMC and an enteric polymer (Eudragit™ L30D). The lag-time in relation to absorption was found to be 7.3 ± 1.2 hours when the thickness of the polymer layer was greatest. The formulation disintegrated in the colon in all six volunteer subjects. Time-controlled formulations have also been prepared using water insoluble ethylcellulose and swellable polymer (HPC). Each of the formulations consisted of a core, drug, swelling agent, and a water-insoluble membrane. The swelling agent HPC absorbed liquid and the ethylcellulose coat disintegrated as the core swelled. A lag time of 4.0 ± 0.5 hours in relation to absorption was found for this formulation in a human bioavailability study, and it was not influenced by food.

A drug delivery system (Pulsincap™), from which there is rapid drug release after a lag-time, has been developed to allow for the release of the drug in the large intestine. The system involves an insoluble capsule body with a hydrogel plug. The plug is ejected from the capsule when it has swelled after a particular lag-time. A release profile is characterized by a period during which there is no release followed by rapid and complete drug release. Release using this system was found to be reproducible *in vitro* and *in vivo*. When gastrointestinal transit of the formulations was followed by means of gammascintigraphy it was found in six of the eight subjects that the device reached the colon before the drug was released. The formulation had been administered with the subjects in a fasting state. Effects of food and gastric retention time were not investigated. In later scintigraphic studies, it was found that the site of release for the drug in the gastrointestinal tract varied. In one subject the formulation even remained in the stomach for a long time, and drug was also released in the stomach. A formulation that involves a plug that erodes rather than a hydrogel plug has also been developed. The aim of these studies were to simplify the Pulsincap™ technology and develop a chronopharmaceutical formulation.

For several decades, 5-fluorouracil (5-FU) has been the only chemotherapeutic agent with clinical activity against colorectal cancer. The 5-FU is an anticancer agent and the most widely used drug in the malignancies arising from breast, gastrointestinal (GI) tract, head, and neck. This antimetabolite antineoplastic agent is widely used for treatment of colorectal. Due to erratic oral bioavailability, intravenous administration of this drug is currently in clinical use. Such an administration mode; however, causes severe GI, hematological, neural, cardiac, and dermatological toxic effects, due to 5-FU cytotoxicity. These are more often encountered with dose intensification strategies, but also occur with moderate doses in adjuvant therapy treatments. Oral site-specific rate-controlled 5-FU delivery is expected to reduce systemic side-effects, and also, to provide an effective and safe therapy for colon cancer with reduced dose and duration of therapy. Colonic drug delivery has gained increased importance not just for the delivery of the drugs for the treatment of local diseases associated with the colon like Crohn's disease, ulcerative colitis, and colon cancer; but also, for the potential that it holds for the systemic delivery of proteins and therapeutic peptides. The large intestine, though difficult to reach by peroral delivery, is still deemed to be the ideal site for the delivery of agents to cure the local diseases of the colon. Due to the distal location of the colon in the GI tract, a colon specific drug delivery system should prevent drug release in the stomach and small intestine, and affect an abrupt onset of drug release upon entry into the colon. Such a system can be formulated utilizing some specific conditions existing in the colon in comparison to other parts of the GI tract. The most critical challenge for such a drug delivery approach is to preserve the formulation during its passage through the stomach and for about the first 6 m of the small intestine. Generally, four primary approaches have been proposed for colon targeted delivery namely pH dependent system, time dependent systems, colonic micro flora activated systems, and pressure controlled systems.

Although oral delivery has become a widely accepted route for the administration of therapeutic drugs, the gastrointestinal tract presents several formidable barriers to drug delivery. The delivery of drugs to the colon has a number of therapeutic implications in the field of drug delivery. In recent times, the colon specific drug delivery system is also gaining importance not only for local drug delivery of drugs but also for the systemic delivery of protein and peptide drugs. The colon, as a site for drug delivery, is also beneficial for the treatment of disease sensitive to circadian rhythms and the delivery of poorly absorbable drugs. The various approaches that can be exploited to target the release of a drug to the colon include prodrug formation, coating with pH sensitive polymer, coating with biodegradable polymers, embedding in biodegradable matrices, hydrogel, time release system, osmotic and bio-adhesive system.

Targeting of drugs specifically for the colon is advantageous in the treatment of diseases such as amoebiasis, Crohn's disease, ulcerative colitis, and colorectal cancer. In addition, it has shown great potential in the oral delivery of therapeutic peptides and proteins, which are unstable in the upper part of the gastrointestinal (GI) tract. The colonic region is recognized as having less diversity and intensity of enzymatic activities than the stomach and the small intestine. Various strategies are available for targeting drug release selectively to the colon. The design of prodrugs is based on the concept of preventing the release of drugs in the stomach and small intestine and drug release is triggered by the use of a specific property at the target site such as altered pH or high activity of certain enzymes in comparison to nontarget tissues. Since it is known that azo function can be reduced in the colon, many novel polymers containing azo groups either in the polymeric backbone or in the cross-links have been synthesized. To promote further selective degradation in the vicinity of the colonic environment, delivery systems have been designed that contain both pH-sensitive acidic monomers and degradable azo aromatic cross-links. Chitosan is a functional linear polymer derived from chitin, the most abundant natural polysaccharide on the earth after cellulose, and it is not digested in the upper GI tract by human digestive enzymes. Chitosan is a copolymer consisting of 2-amino-2-deoxy-D-glucose and 2-acetamido-2-deoxy-D-glucose units links with β-(1-4)bonds. It should be susceptible to glycosidic hydrolysis by microbial enzymes in the colon because it possesses glycosidic linkages similar to those of other enzymatically depolymerized polysaccharides. The polysaccharide, on reaching the colon, undergoes assimilation by microorganisms or degradation by enzymes or break down of the polymer back bone. Thus, leading to a subsequent reduction in molecular weight and thereby loss of mechanical strength and unable to hold the drug any longer. Chitosan has drawn attention for its potential to achieve site-specific delivery to the colon. Chitosan is a natural linear polyamine with a high ratio of glucosamine to acetyl-glucosamine units, and a weak base that carries a positive charge. Its solubility is pH-dependent, and it reacts readily with negatively charged surfaces and materials, including polymers and DNA. Ionic gelation, complex coacervation, emulsion cross-linking, and spray-drying are methods commonly used for the preparation of chitosan nanoparticles. Among those methods, ionic gelation and complex coacervation are mild processes occurring in a pure aqueous environment and are ideal for maintaining the in-process stability of drugs. Ionic gelation and complex coacervation are very similar except that the former involves the gelation of chitosan using an electrolyte such as tripolyphosphate (TPP), whereas the latter employs an oppositely charged ionic polymer such as alginate. This new type of chitosan (CS) hydrogel uses dextran sulfate (DS) as a polyanionic polymer. It was developed to achieve complex coacervation for the incorporation and controlled release of an antiangiogenesis hexapeptide. It is the first report in describing the use of DS to formulate CS-based hydrogels. Although there have been investigations of how the properties of CS and formulation variables such as CS molecular weight (MW), concentrations of CS and 5-ASA, and formulation pH affect the formation and encapsulation capability of hydrogels; to our knowledge, no attempts have been made to study how the ratio of CS to the oppositely charged polymer influences the formulation and properties of hydrogel. In this study, 5-aminosalicylic acid (5-ASA) was chosen as a model drug. These hydrogels had good spherical geometry. It is obvious that the surface of the hydrogel shrank and a densely cross-linked gel structure was formed. This may explain the greater retardation of the drug release in matrices with higher crosslink content. The average drug entrapment was found to be 81.21%±1.86% in the hydrogel.

Figure 2. Morphology of chitosan-dextran sulfate (CS-DS) hydrogel by emission scanning electron microscopy field.

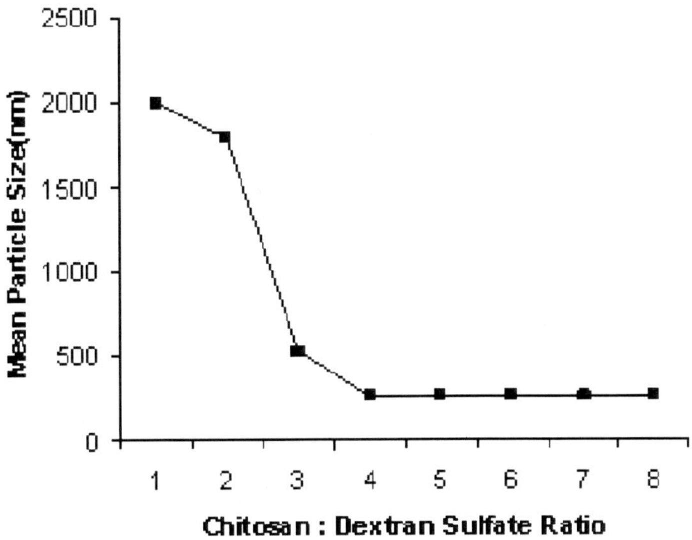


Figure 3. Influence of the Chitosan: Dextran sulfate ratio on the particle size.

Results of drug content and EE demonstrated that the drug content increased from 10.23±0.50 mg/100mg to 26.24±0.43 mg/100mg of hydrogel which also increased the amount of drug from 5% to 20% wt/wt. No significant increase in the drug content was observed further when the amount of the drug increased, i.e., above 15% wt/wt, which could be due to the limited solubility of the drug in DMF. This is endorsed from the presence of drug particles on the surface of the hydrogel prepared with 20% of the drug concentration. The percent of EE was increased up to 86.32%±0.20% along with increasing polymer concentrations to 4%. Concentration of the cross-linking agent exhibited no significant effect on the percent of EE. FESEM images of super paramagnetic CS-DS hydrogel (Figure 2) showed that hydrogel had a solid and near consistent structure.

Furthermore, the incorporation of 5-ASA into the hydrogel produced a smooth surface and compact structure. The particle size observed in FESEM is smaller than that measured by the Zetasizer. This is because dried hydrogel was used in FESEM, whereas particles in the liquid dispersion were analyzed by the Zetasizer. CS-DS hydrogel are hydrophilic and would be expected to swell in water, thus producing a large hydrodynamic size when measured by the Zetasizer (Figure 3).

A 5-aminosalicylic acid (5-ASA) – loaded hydrogel was obtained spontaneously upon the mixing the DS aqueous solution (0.1%wt/vol) with the CS solution (0.1% wt/vol) under magnetic stirring, with 5-ASA dissolved in CS-DS solution. The incorporation of 5-ASA into the CS-DS hydrogel resulted in a sharp increase in the particle size of the nanoparticle dispersion. The significant increases in particle size give a good induction of the incorporation of 5-ASA into CS-DS hydrogel. A study was undertaken to investigate the effect of the order of 5-ASA mixing with CS and DS. The data obtained show that the order of 5-ASA mixing had no effect on the size, entrapment efficiency, and yield of 5-ASA-loaded hydrogel. The effect of drug concentration, chitosan-dextran sulfate concentration, GST concentration, and cross-linking time on *in vitro* drug release was also observed. *In vitro* drug release after 5 hours was 86.3%±4.0% in the case of hydrogel having 15% drug, while it was

88.2%±3.1% for hydrogel with 20% drug. The effect of chitosan-dextran sulfate on the drug release was found to be meager. It is also observed that the amount of the drug released from hydrogel decreased when increasing the cross-linking time. These properties are probably explained by the promotion of cross-links between chitosan-dextran sulfate chains and GST. Freeze-drying the samples resulted in larger and more porous hydrogel compared with air-dried hydrogel. Freeze-drying had the advantage of avoiding drug extraction by immediately freezing and removing the water present within the hydrogel and a burst effect during the dissolution study (Figure 4).

Conventional dissolution testing is less likely to accurately predict *in vivo* performance of colon delivery systems triggered by bacteria residing in the colon (because aspects of the colon's environment, *i.e.*, scarcity of fluid, reduced motility, and presence of microflora, cannot be simulated in conventional dissolution methods). Hence, release studies were performed in an alternate release medium.

The release profile of 5-ASA-loaded hydrogel was evaluated in water and a phosphate buffer, which was either in a different ionic strength or with saline, to study the underpinning mechanisms for drug release. The greatest release of 5-ASA-loaded hydrogel occurred in the release media of a high ionic strength. In contrast, a significantly small portion of 5-ASA was released in water over the release study period. The burst release was observed with hydrogel, and this may have arisen due to desorption of loosely attached 5-ASA from the surface of the matrix polymers (Figure 5).

The opposite charge found in the CS-DS hydrogel was responsible for the formation of nanoparticles. The charge ration between the negatively charged sulfate groups (N) in DS and the positively charged amine groups (P) in CS was determined. Under the experimental conditions(pH=3-4), DS carries ~74 sulfate groups per mole, equivalent to $5.78 \times 10^{-3}$ negatively charged groups per gram of DS; CS has ~2073 amino groups per gram of CS.

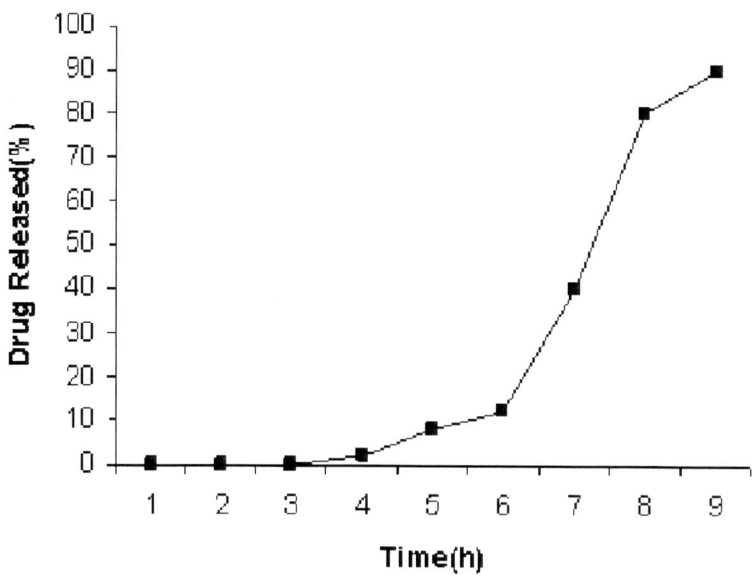


Figure 4. The in vitro release profile of 5-aminosalicylic acid from chitosan-dextran sulfate hydrogel in various simulated gastrointestinal fluids (n=3).

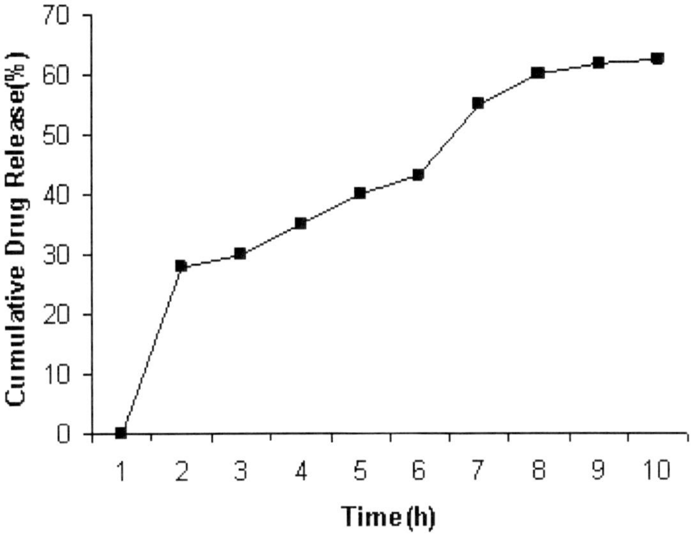


Figure 5. *In Vitro* Drug Release profiles for Chitosan-Dextran Sulfate Hydrogel in SCF (pH 7.0) involving the Presence of 5-ASA.

Table 2 shows conductivity of dispersion for the CS-DS hydrogel. Natural polymers have potential pharmaceutical applications because of their low toxicity, biocompatibility, and excellent biodegradability. In recent years, biodegradable polymeric systems have gained importance in the design of surgical devices, artificial organs, for drug delivery systems with different routes of administration, carriers of immobilized enzymes and cells, biosensors, ocular inserts, and materials for orthopedic applications. These polymers are classified as either synthetic (like polyesters, polyamides, and polyanhydrides) or natural (like polyamino acids and polysaccharides). Polysaccharide-based polymers represent a major class of biomaterials, which include agarose, alginate, carageenan, dextran, and chitosan. Chitosan [β(1,4)2-amino-2-D-glucose] is a cationic biopolymer produced by alkaline N-deacetylation of chitin, which is the main component found in the shells of crab, shrimp, and krill. Chitosan is a functional linear polymer derived from chitin, the most abundant natural polysaccharide on the earth after cellulose, and it is not digested in the upper GI tract by human digestive enzymes. Chitosan is a copolymer consisting of 2-amino-2-deoxy-D-glucose and 2-acetamido-2-deoxy-D-glucose units links with β-(1-4)bonds. It should be susceptible to glycosidic hydrolysis by microbial enzymes in the colon because it possesses glycosidic linkages similar to those of other enzymatically depolymerized polysaccharides. Among diverse approaches that are possible for modifying polysaccharides, grafting of synthetic polymer is a convenient method for adding new properties to a polysaccharide with minimum loss of its initial properties. Graft copolymerization of vinyl monomers onto polysaccharides using free radical initiators, has attracted the interest of many scientists. Up to now, considerable works have been devoted to the grafting of vinyl monomers onto the substrates, especially starch and cellulose. Existence of polar functionally groups as carboxylic acid need not only be for bioadhesive properties but also for pH-sensitive properties of a polymer. The increase of MAA content in the hydrogels provides more hydrogen bonds at low pH and more electrostatic repulsion at high pH. A part of our research program is chitosan modification to prepare materials with pH-sensitive properties for uses as drug delivery. The free radical graft copolymerization polymethacrylic acid onto Chitosan was carried out at 70°C, bis-acrylamide as a cross-linking agent and persulfate as an initiator. Polymer bonded drug usually contain one solid drug bonded together in a matrix of a solid polymeric binder. They can be produced by polymerizing a monomer such as methacrylic acid (MAA), mixed with a particulate drug, by means of a chemical polymerization catalyst, such as AIBN or by means of high-energy radiation, such as X-rays or γ-rays. The modified hydrogel and satranidazole as a model drug was converted to nanoparticles by freeze-drying method. The equilibrium swelling studies and *in vitro* release profiles were carried out in enzyme-free simulated gastric and intestinal fluids [SGF (pH 1) and SIF (pH 7.4), respectively)]. The influences of different factors, such as content of MAA in the feed monomer and swelling were studied. In the present study, the beads were prepared by the classical method, which involves spreading a uniform layer of polymer dispersion followed by a drying step for removal of solvent system. Since the methodology of bead preparation involved a heating step, it may have had a detrimental effect on the chemical stability of drug. Hence, the stability assessment of satranidazole impregnated in the bead was done using a stability-indicating method. For this purpose, satranidazole was extracted from the bead and analyzed by HPLC. A single peak at 21.2 minutes representing satranidazole (with no additional peaks) was detected in the chromatogram, suggesting that the molecule was stable during preparation of the beads.

Satranidazole was extracted from different regions of N-sulfonato-N,O-carboxymethylchitosan/PMAA bead using acetonitrile: ETOH(1:1 vol/vol) solvent system. After normalization of satranidazole concentration on the basis of bead weight, the variation in distribution for satranidazole in different regions of the bead was <16%. The composition of the polymer defines its nature as a neutral or ionic network can further determine its hydrophilic/hydrophobic characteristics. Ionic hydrogels, which could be cationic, containing basic functional groups or anionic, containing acidic functional groups, have been reported to be very sensitive to changes in the environmental pH. The swelling properties of the ionic hydrogels are unique due to the ionization of their dependent functional groups. The equilibrium swelling behavior of ionic hydrogels containing acidic and/or basic functional groups is illustrated in Figure 6.

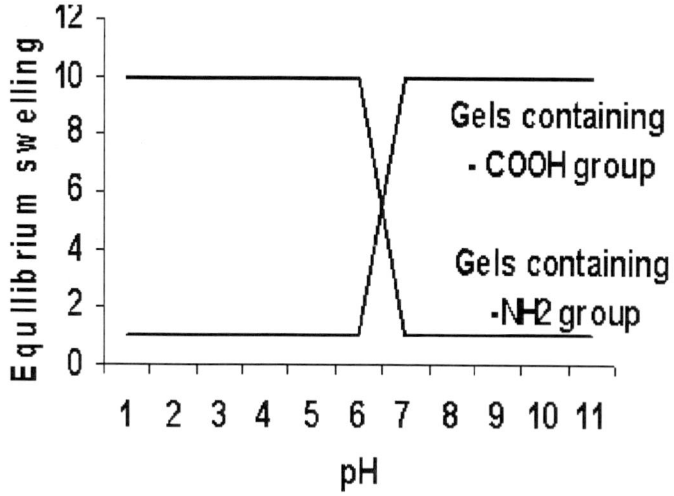


Figure 6. Equilibrium swelling results in response to pH changes.

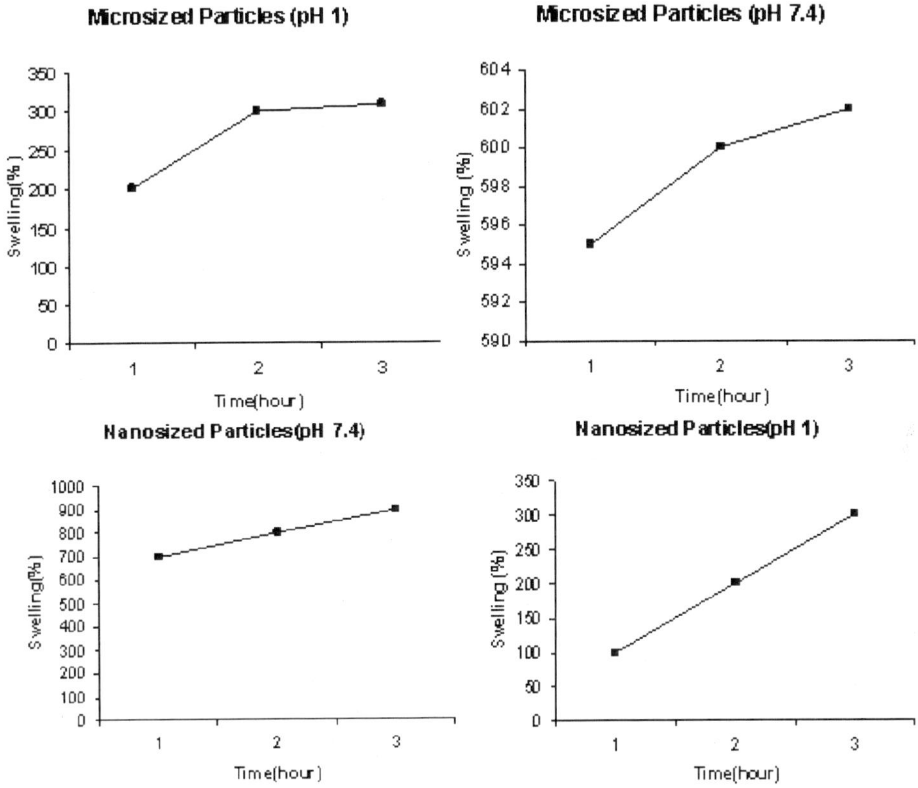


Figure 7. Time-dependent swelling behaviors of micro and nano carriers for the satranidazole drug model with a time function at 37ºC.

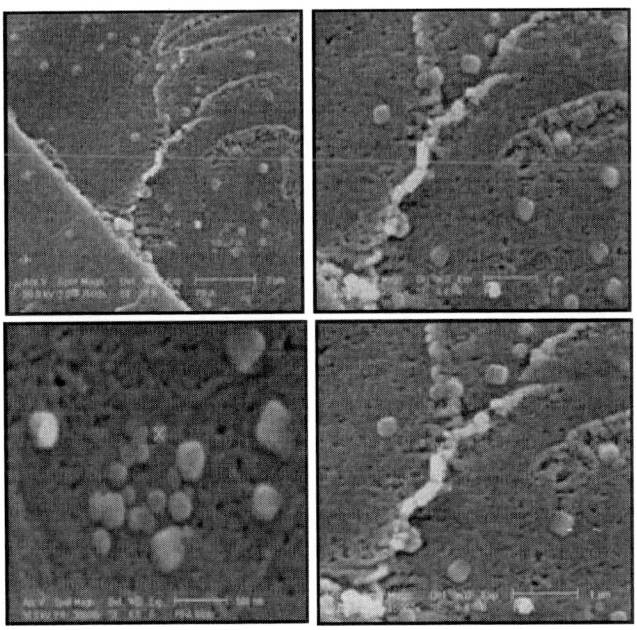

Figure 8. Scan (SEM) of NOCCS-Polymethacrylic acid hydrogel beads with satranidazole.

Hydrogels containing basic functional groups is found to have increased swelling activity in acidic conditions and reduced for basic conditions. The pH sensitive anionic hydrogels shows low swelling activity in acidic medium and very high activity in basic medium. As shown in Figure 7, an increase in the content of MAA in the feed monomer mixtures resulted in less swelling in simulated gastric fluid but greater swelling in and simulated intestinal fluids.

This is because the increase of MAA content in the hydrogels provides more hydrogen bonds at low pH and more electrostatic repulsion at high pH. Figure 8 shows the scanning electron microscope (SEM) of graft NOCCS copolymer with polymethacrylic acid and nano polymer bonded drug, respectively.

Nano and micro polymer bonded drugs (50 mg) were poured into 3 mL of aqueous buffer solution (SGF: pH 1 or SIF: pH 7.4) (Figure 9). The mixture was introduced into a cellophane membrane dialysis bag. The bag was closed and transferred to a flask containing 20 mL of the same solution maintained at 37° C. The external solution was continuously stirred, and 3 mL samples were removed at selected intervals. The removed volume was replaced with SGF or SIF. The triplicate samples were analyzed by UV spectrophotometer, and the quantity of satranidazole was determined using a standard calibration curve obtained under the same conditions.

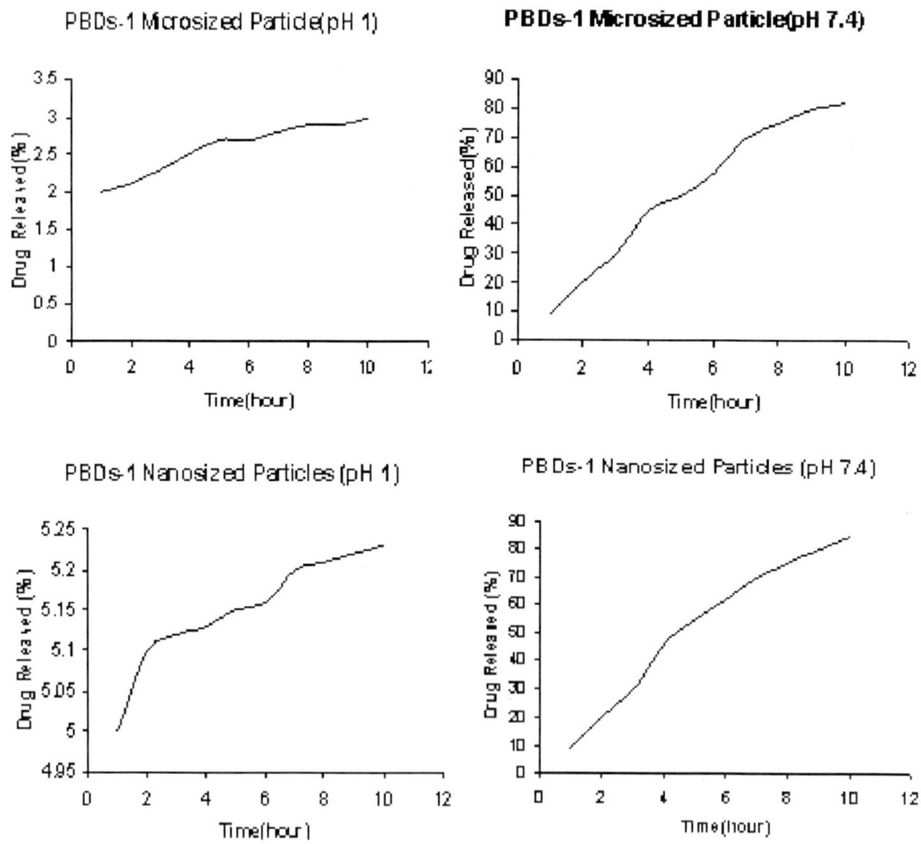


Figure 9. Graphs that indicate the release of the satranidazole drug from micro and nano polymeric carriers with a time function at 37°C.

**Table 1. DSC data and composition of copolymer**

| Polymer Samples | Molar composition of monomers in the feed | | | | Degree of Substitution (DS) | $T_g$ (° C) |
|---|---|---|---|---|---|---|
| | NOCCS (gr) | MAA (gr) | CA (gr) | IN (gr) | | |
| P- 1 | 1 | 3 | 0.05 | 0.05 | 0.52 | 135 |
| P- 2 | 1 | 2 | 0.05 | 0.05 | 0.49 | 142 |

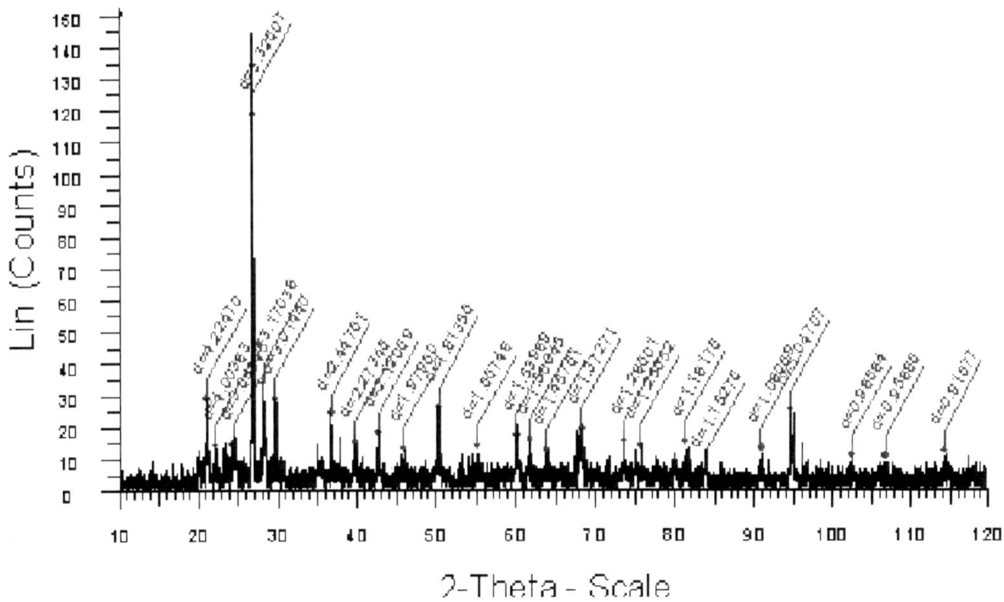


Figure 10. XRD patterns of NOCCS-Polymethacrylic acid hydrogel beads.

The thermal behavior of a polymer is important for controlling the release rate in order to have a suitable drug dosage form. The glass transition temperature (Tg) was determined from the DSC thermo grams. The values are given in Table 1. The higher Tg values probably related to the introduction of the cross-links, which would decrease the flexibility of the chains and the ability of the chains to undergo segmental motion, which would increase the Tg values. On the other hand, the introduction of a strongly polar carboxylic acid group can increase the Tg value because of the formation of internal hydrogen bonds between the polymer chains.

X-ray diffraction is a proven tool to study crystal lattice arrangements and it yields very useful information on the degree of crystallinity. The X-ray diffraction pattern of hydrogel with satranidazole had several high-angle beads at the following 2θ values: 21, 22, 25, 26.5, 28, and 32. The 26.5 ° 2θ peak had the highest intensity, and the hump in the baseline occurred from 7 ° to 45 ° 2θ, as observed for the chitosan bead (Figure 10).

In order to develop a local delivery system for satranidazole, a biodegradable N-sulfonato-N, O-carboxymethylchitosan-PMAA bead was attempted, which to our knowledge is the first effort of its kind. The potential formulation problems were anticipated since chitosan is only soluble in aqueous acidic solutions; whereas satranidazole, being a hydrophobic drug, is insoluble under similar conditions. In the early stages of formula optimization studies, the satranidazole was incorporated into a bead. The primary mechanisms used in the matrix systems for the release of drugs for *in vitro* were swelling, diffusion, and disintegration. *In vitro* degradation of chitosan beads were prepared by a solution casing method which occurred less rapidly, to the degree that 73% deacetylated showed slower biodegradation. Since the grade of chitosan used in the present study was of high molecular weight with a degree of deacetylation ≥85%, significant retardation for the release of satranidazole from the bead is attributed to the polymer characteristics. In addition, diffusion of satranidazole may have been hindered by the increased tortuosity of the polymer accompanied by a swelling mechanism. As observed from SEM photomicrographs, the crystals of satranidazole have a different appearance than recrystallized satranidazole. These nanoparticles do not have clearly defined crystal morphological features in the SEM photomicrographs. Hence, it appears that the irregularly shaped particles are deposited on the surface with the poloxamer, which gives them an appearance resembling that of coated particles. X-ray diffraction technique is also used to study the degree of crystallinity of pharmaceutical drugs and excipients. A lower 2θ value indicates larger d-spacings, while an increase in the number of high-angle reflections indicates higher molecular state order. In addition, broadness of reflections, high noise, and low peak intensities are characteristic of a poorly crystalline material. A broad hump in the diffraction pattern of chitosan extending over a large range of 2θ suggests that chitosan is present in an amorphous state in the bead. Existence of polar functionally groups as carboxylic acid need not only be for bioadhesive properties but also for pH-sensitive properties of polymer. Because the increase of MAA content in the hydrogels provides more hydrogen bonds at low pH and more electrostatic repulsion at high pH. This is part of our research program on chitosan modification to prepare materials with pH-sensitive properties for uses in drug delivery. For this novel research, the absorption of paclitaxel could be enhanced with the administration of chitosan-polymethacrylic acid nanoparticles because of their greater permeability properties. In the present study, nanoparticles were prepared by the classical method, which involves spreading a uniform layer of polymer dispersion followed by a drying step for the removal of the solvent system. Since bead preparation methodology involved a heating step, it may have had a detrimental effect on the chemical stability of the drug. Hence, stability assessment of paclitaxel impregnated in bead was done using a stability-indicating method. For this purpose, paclitaxel was extracted from bead and analyzed by HPLC. A single peak at 21.2 minutes representing paclitaxel (with no additional peaks) was detected in the chromatogram, suggesting that the molecule was stable during preparation of beads. In order to develop a local delivery system for paclitaxel, a biodegradable N-chitosan-PMAA bead was attempted, which to our knowledge is the first effort of its kind. Potential formulation problems were anticipated since chitosan is only soluble in aqueous acidic solutions; whereas paclitaxel, being a hydrophobic drug, is insoluble under similar conditions. In the early stages of formula optimization studies, it was observed that paclitaxel was incorporated into a bead. The primary mechanisms used in the matrix systems for the release of drugs for *in vitro* were swelling, diffusion, and disintegration. *In vitro* degradation of chitosan-PMAA nanoparticles prepared by solution casing method occurred less rapidly to the degree that 73% deacetylated showed slower biodegradation. Since the grade of chitosan used in the present study was of high molecular weight with a degree of deacetylation ≥85%, significant retardation for release of paclitaxel from nanoparticles is attributed to the polymer characteristics. In addition, diffusion of paclitaxel may have been hindered by the increased tortuosity of the polymer accompanied by a swelling mechanism. As observed from SEM photomicrographs, the crystals of paclitaxel have a different appearance than recrystallized paclitaxel. These nanoparticles do not have clearly defined crystal morphological features in the SEM photomicrographs. Hence, it appears that the irregularly shaped particles are deposited on the surface with a poloxamer, which gives them an appearance resembling that of coated particles. The X-ray diffraction is also used to study the degree of crystallinity on pharmaceutical drugs and excipients. A lower 2θ value indicates larger d-spacings, while an increase in the number of high-angle reflections indicates a higher molecular state order. In addition, broadness of reflections, high noise, and low peak intensities are characteristic of a poorly crystalline material. A broad hump in the diffraction pattern of chitosan extending over a larger range of 2θ suggests that nanoparticles in the chitosan are present in an amorphous state. Recent studies have been reported with an investigation on the effect of 4.0 generation (G) PAMAM dendrimer present in an anionic phospholipids composition, consisting of hydrogenated soyaphosphotidyl-choline, cholesterol, diacetyl phosphate and poly(ethylene glycol)-derivatized phosphotidylethanolamine, on the hydration and liquid crystalline structure formation. In this novel study, the 4.0 generation PAMAM dendrimer was exhaustively studied as the controlled-release system for the parenteral administration with the model drug 5-aminosalicylic acid and then analyzed using various release kinetic studies. This study gives us an insight about the biodistribution drug pattern and localization at the site of inflammation with a PAMAM dendrimer. In the present study, films were prepared by the classical method, which involves spreading a uniform layer of polymer dispersion that is followed by a drying step in the removal of the solvent system. Since film preparation methodology involved a heating step, it may have had a detrimental effect on the chemical stability of drug. Hence, the stability assessment of 5-ASA impregnated in film was done using a stability-indicating method. For this purpose, the 5-ASA was extracted from film. The 5-ASA was extracted from different regions of Ac-PAMAM-CS film using acetonitrile: ETOH (1:1 vol/vol) solvent system. After normalization, observations were taken for the amount of 5-ASA weight on the coated film. These results indicated that the distribution variation for 5-ASA in different regions of film was <18%. Mechanical strength for these films is described in terms of tensile strength, and brittle films are characterized by a decrease in the percentage of elongation at the break point. The area under the curve is related to the energy required to break this polymeric material, and tough polymers have larger areas which require a greater amount of energy for rupture. In order to understand the arrangement of polymer chains in the presence of 5-ASA, force-time profiles of films were generated as shown in Figure 11.

Although the force of elongation at the break point is slightly lower on the film in the presence of 5-ASA, the area under the profile had increased. For convenience of interpretation, each profile is further described in terms of ascending and descending portions.

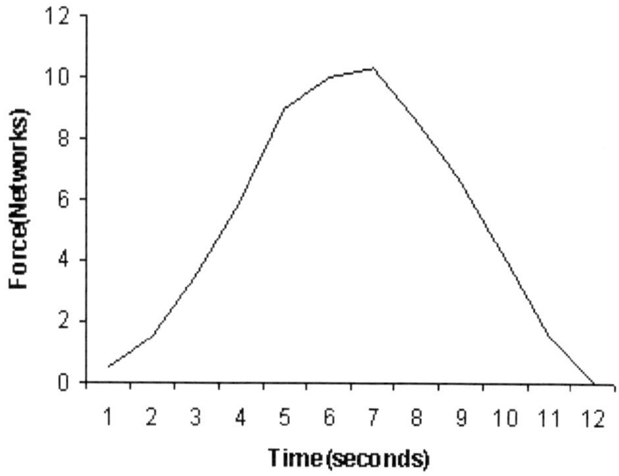


Figure 11. Force-time profiles of hydrogel film to study the influence of formulation approach on mechanical properties of film.

The plateau time for the ascending portion of 5-ASA–Ac-PAMAM-CS film was greater in comparison to the control film. The brittleness is reflected by the plateau, where the longer time is indicative to the lack of brittleness in the film. In order to confirm *in vivo* degradation of the film, a time-dependent visual monitoring of film texture and integrity following the implantation in mice was performed. On days 2,10,20, and 50, mice were humanely killed by cervical dislocation, and an incision was made in the site for the film examination. After 2 days of implantation, it was observed that film turned white and opaque when compared with the control film, which was initially yellowish and translucent. During the progression of implantation period, the film turned soft and delicate (on the $10^{th}$ day) with loss of mechanical strength. In a period of 30 to 40 days, the film commenced to embed into the surrounding tissues. After 50 days the film was completely embedded. Although this experiment was qualitative in nature, the film's loss of integrity and identity over time confirms that the implant was amenable to biodegradation. The tissue responses in the implanted film were studied by a microscopic tissue examination in the implanted area (Figure 12).

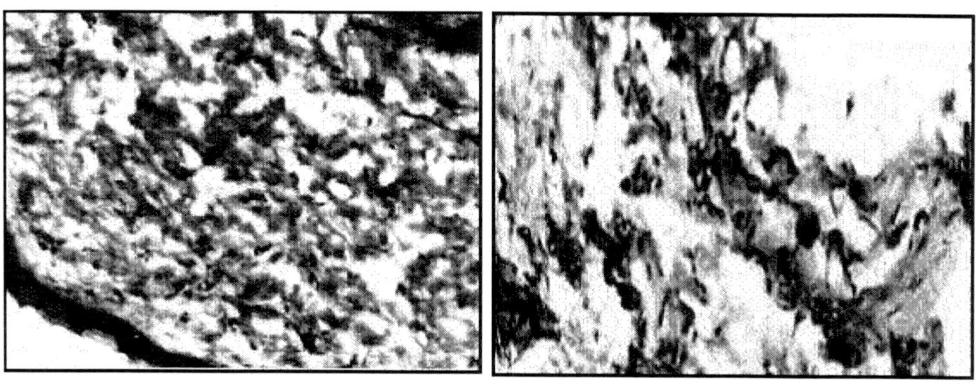

Figure 12. Histopathological examination for generalized inflammatory response to assess biocompatibility of subcutaneous implanted hydrogel film in mice.

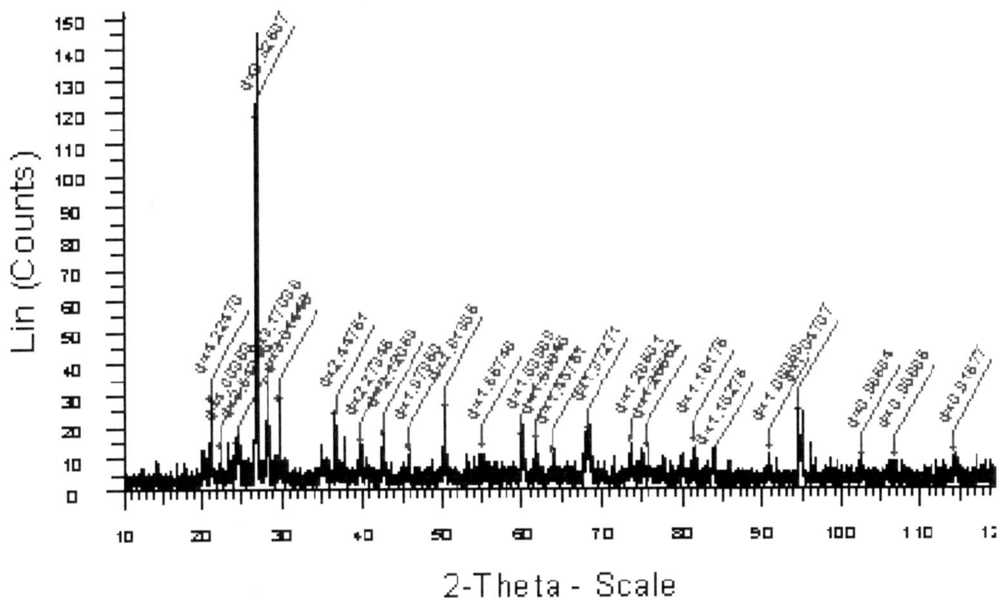


Figure 13. X-ray diffraction patterns of hydrogel film at ambient temperature to examine solid-state features of 5-aminosalicylic acid in film.

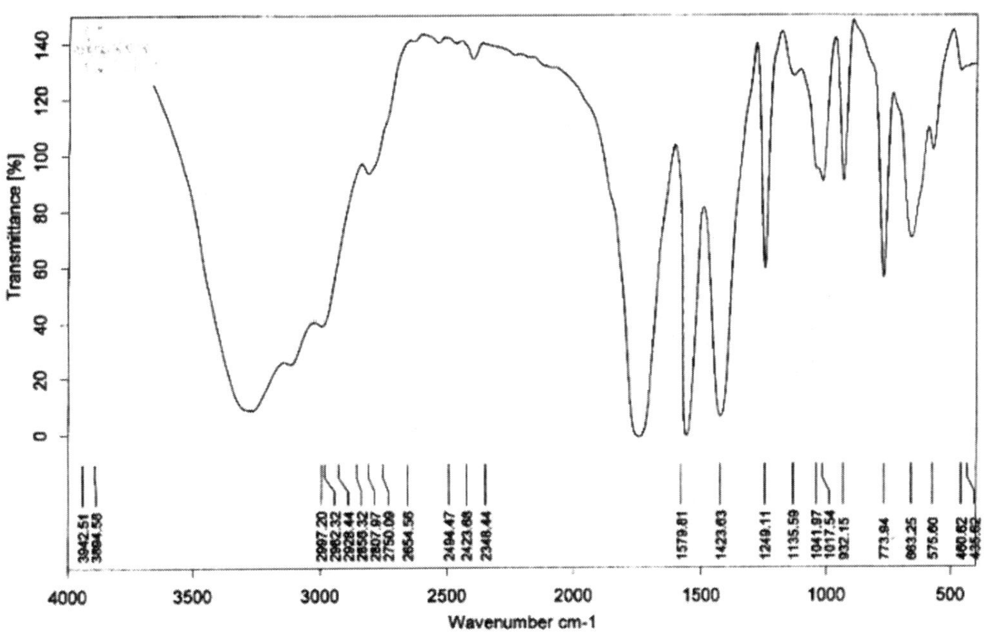


Figure 14. FT-IR spectra of Hydrogel film to examine the alterations in molecular vibrations of 5-aminosalicylic acid.

The lack of increase in macrophages at the site of implantation suggests that inflammatory responses were either minimal or absent. It had been reported previously that the degree of deacetylation for CS is a key factor in determining not only the release rate but also inflammatory responses. Typically, CS gels with low degrees of deacetylation have short residence times and higher degrees have longer residence time with minimal or no inflammation. In the present study, the grade of Ac-PAMAM-CS used for the film fabrication was ≥85% deacetylated, hence, the high degree of deacetylation of CS may be the reason for reducing local inflammatory responses following implantation. X-ray diffraction is a proven tool to study crystal lattice arrangements and yields very useful information on the degree of sample crystallinity. X-ray diffraction pattern of 5-ASA-Ac-PAMAM-CS hydrogel film were obtained and compared, which revealed clear differences in the molecular state of 5-ASA (Figure 13). X-diffractogram of 5-ASA-Ac-PAMAM-CS hydrogel film had shown several high-angle diffraction peaks that were observed at the following 2θ values: 21.3°, 22.5°, 24.1°, 27.4°, 28.2°, 37.4°, 43.3°, and 51.6°. The 27.4° 2θ peak had the highest intensity observed for hydrogel film. Transmission in the infrared spectra of hydrogel powder was acquired to draw information on the molecular state of hydrogel (Figure 14). CS is an amino glucose characterized by a small proportion of the amide group via an amide linkage with acetic acid. In the infrared spectrum, powder CS hydrogel exhibited a broad spectrum at 3431cm$^{-1}$, which is assigned to the N–H and hydrogen bonded O–H strength vibration frequencies, while a sharp peak at 3600 cm$^{-1}$ is due to the free O–H bond stretch of glucopyranose units. Further, in the C–H strength region of FT-IR spectrum, the higher intensity peak at 2920 cm$^{-1}$ is assigned to the asymmetric and the lower intensity peak at 2835 cm$^{-1}$ is assigned to the symmetric modes of $CH_2$. In addition, the characteristic band due to $CH_2$ scissoring, which usually occurs at 1423 cm$^{-1}$ was also present in the sample. Since the grade of CS hydrogel used in the present study was ≥85% deactylated, an amide bond peak was present in the spectra and the C=O strength of amide bond was observed at 1679 cm$^{-1}$. The effect of drug concentration, CS-Ac-PAMAM concentration, GST concentration, and cross-linking time on *in vitro* drug release was also observed. *In vitro* drug release after 5 hours was 86.3%±4.0%; in the case of hydrogel having 15% drug, while it was 89.2%±3.1% for hydrogel with 20% drug. The effect of CS- Ac-PAMAM dendrimer on the release of drug was found to be meager. It is also observed that the amount of released drug from hydrogel decreased with an increased cross-linking time (Figure 15).

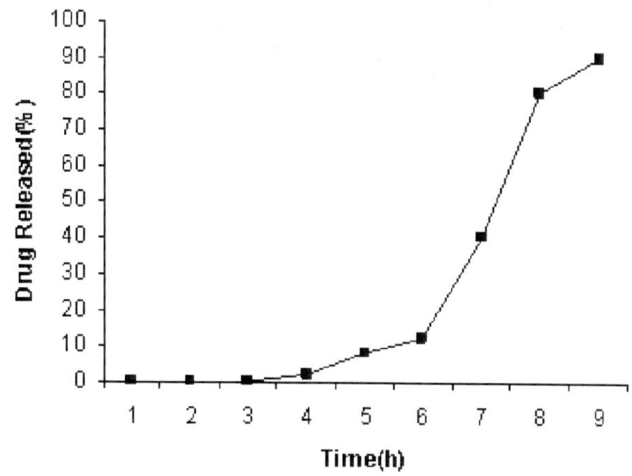


Figure 15. The in vitro release profile of 5-aminosalicylic acid from chitosan-PAMAM hydrogel in various simulated gastrointestinal fluids (n=3).

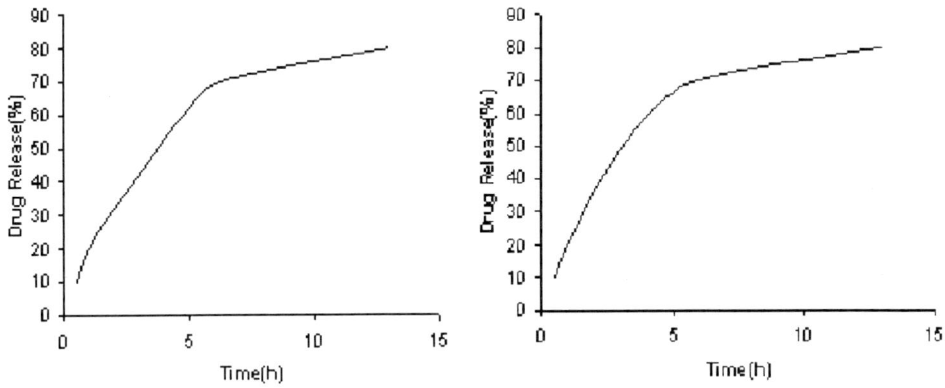


Figure 16. The release of drug (5-ASA) from (a) micro; and, (b) nano polymeric carriers with time function at 37ºC.

The concentration of 5-ASA in the decanted Tris HCl buffer and 2 washing solutions was determined by measuring the absorbance at 235 nm using a GBS Cintra 10-UV-Visible Spectrophotometer (Shimadzu, Japan). The determinations were made in triplicate, and results were averaged (Figure 16).

The most important characteristic of any drug is efficiency. This characteristic may often be reduced because of the inability to deliver the drug to the specific cells or tissues. After administration, the drug may pass through different physiologic barriers and /or pathways, decreasing the actual amount drug that reaches the site. In the search for an ideal carrier system, the dendrimers may have significant potential. Dendrimers are synthetic macromolecules with a well-defined globular structure. The need for advanced materials with improved and new properties for a variety of technological applications has created a demand for both new forms of matter and for polymers that have highly controlled molecular architectures. The established approach to dendritic macromolecules has traditionally involved a divergent process in which growth is started from a polyfunctional core and continued outwards in a stepwise manner that affords larger and larger macromolecules as the process is continued. The fundamental attribute of the convergent approach is that it begins at what will be the periphery of the molecule, proceeding inwards. It is this feature, more than any other, that allows for unparalleled control over molecular architecture. The potential of dendrimers and hyper-branched polymers as a drug carrier system were investigated using poly (2-hydroxy ethyl methacrylate), since the evaluation in methodologies for cellular activity of dendrimer are well known.

Also explored was the interaction of the poly(2-hydroxy ethyl methacrylate) with aromatic polyetheric dendrimer. The nature of the interaction was characterized by FT-IR and $^{1}$H-NMR spectroscopy.

Figure 17 shows the FT-IR spectrum of the G2-$COOCH_3$ dendrimer(AB8, heptamer) where the percentage of transmittance is plotted as a function of a wave number ($cm^{-1}$). The characteristic FT-IR peaks at 3078, 3053, 2961, 1718 $cm^{-1}$ are due to the presence of =CH bond stretching vibrations, the aliphatic CH bond in protected groups (t-butyldimethylsilyloxy) and carbonyl (C=O) group, respectively.

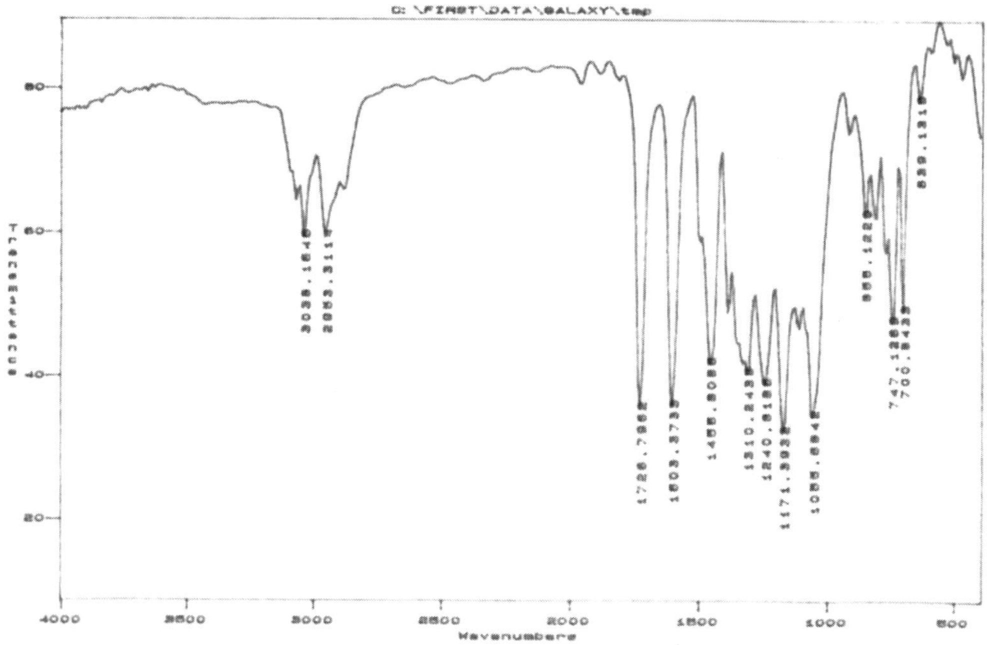


Figure 17. The FT-IR spectrum plotted for the [G2] –COOCH3 Dendrimer.

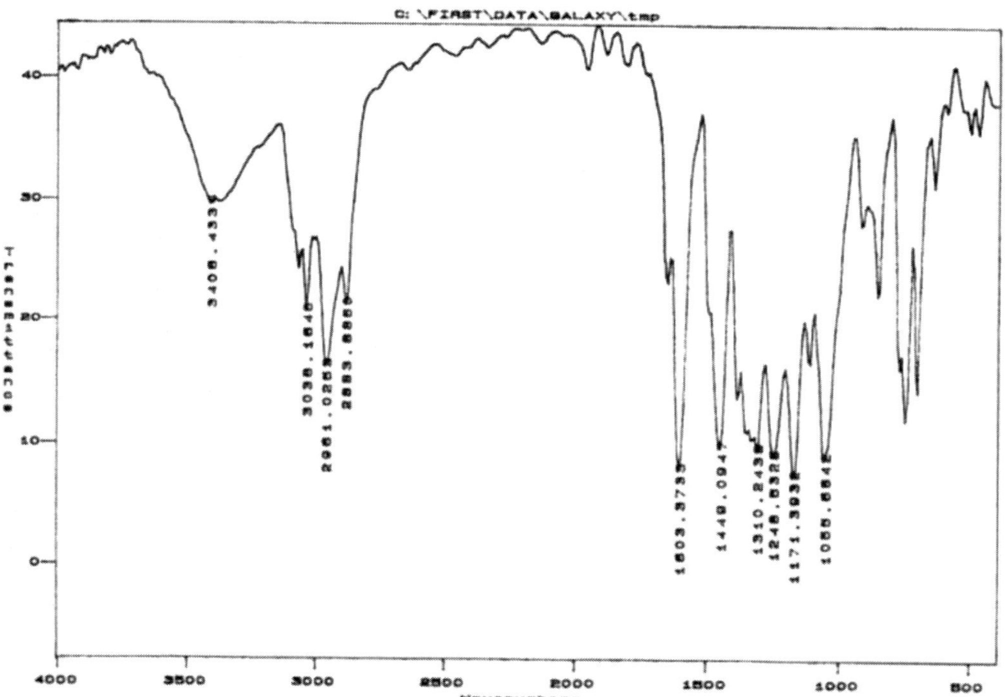


Figure 18. The FT-IR spectrum plotted for the [G2] – CH2OH Dendrimer.

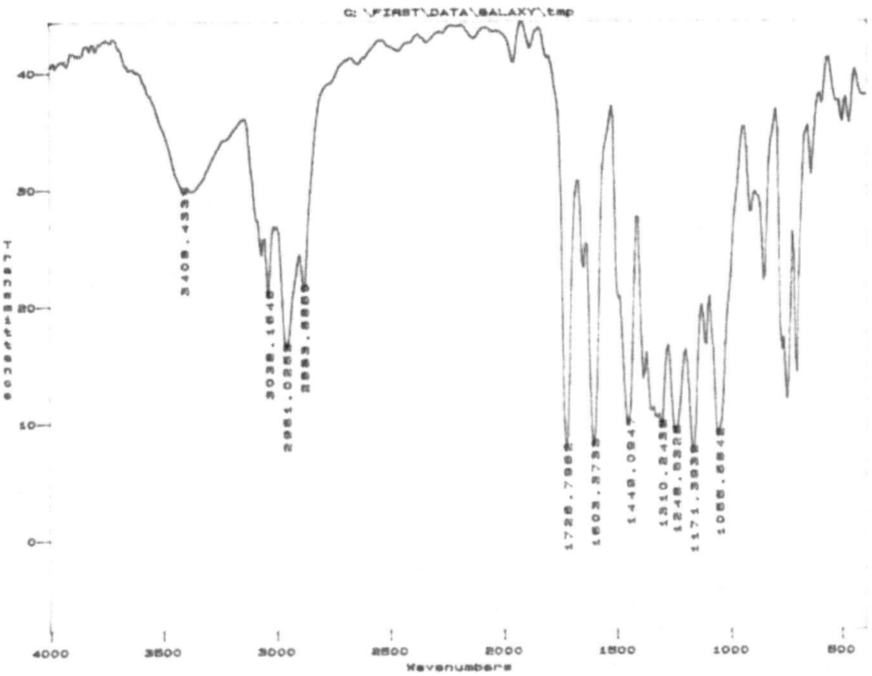


Figure 19. FT-IR spectrum plotted for [G2] – CH2OH Dendrimer/poly(2-hydroxy ethyl methacrylate)/5-ASA nanocomposite.

Figure 18 shows the FT-IR spectrum of the G2-$CH_2OH$ dendrimer (AB8, heptamer). The characteristic FT-IR peaks at 3408, 2960, 2932, 2860, 1544 $cm^{-1}$ are due to the presence of OH phenolic group, the aliphatic and aromatic =CH bond in protected methyl groups and C=C bond of aromatic group, respectively.

Also, Figure 19 shows the FT-IR spectrum of the G2-$CH_2OH$ dendrimer (AB8, heptamer)/ poly(2-hydroxy ethyl methacrylate) nanocomposite. The characteristic FT-IR peaks at 3408, 2960, 2932, 2860, 1544, 1726$cm^{-1}$ are due to the presence of OH phenolic group, the aliphatic and aromatic =CH bond in protected methyl groups and C=C bond of aromatic group and carbonyl (C=O) group of 2-hydroxy ethyl methacrylate, respectively.

Figure 20 shows the SEM of aromatic polyether dendrimer / poly (2-hydroxy ethyl methacrylate) nanocomposite with 5-aminosalicylic acid (5-ASA) as the drug model. This nanocomposite is very sensitive to temperature due to the interaction of electrons. Scanning electron micrography images were obtained from a diluted solution of the nanocomposite particles. The white particles were 5-ASA nanoparticles. The SEM image revealed the presence of 5-aminosalicylic acid (5-ASA) particles in poly-functional dendrimeric matrix, which were distributed throughout the nanocomposite, and were also confirmed by $^1$H-NMR studies.

The ability of the dendrimer to form a complex with 2-hydroxy ethyl methacrylate monomer depends on the core-surface groups of dendrimer, electrostatic interactions between the dendrimer and the monomer, and the ability of the drug to form a conjugate with the dendrimer through chemical bonding.

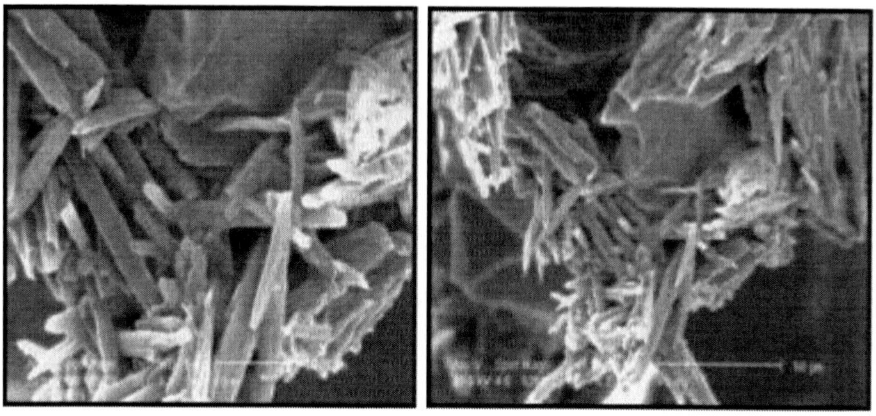

Figure 20. The drug model scans of [G2] – CH2OH Dendrimer/2-hydroxy ethyl methacrylate/5-ASA nanocomposite.

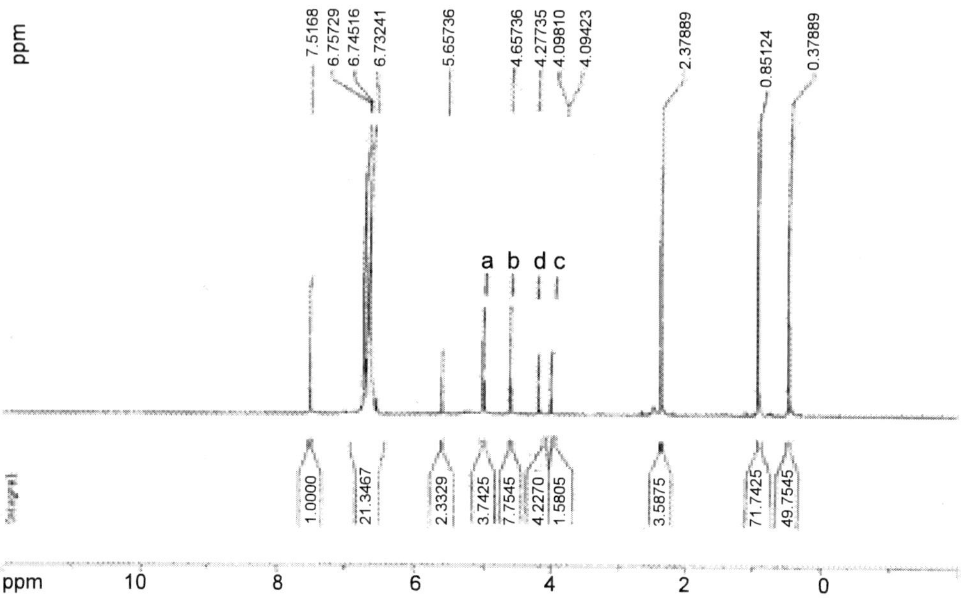


Figure 21.Three regions are shown in the 400 MHz 1H-NMR spectrum for the polymer [G2] – CH2OH Dendrimer/2-hydroxy ethyl methacrylate conjugate(CDCl3).

Therefore, it is possible to manipulate the incorporation process for a given monomer by the selection of the appropriate dendrimer and surface functionality. One might expect that the 2-hydroxy ethyl methacrylate with the carboxylic group may form a complex with the surface OH groups of polyether dendrimer. Figure 21 shows the 400 MHz $^1$H-NMR spectrum of [G2]-$CH_2OH$ Dendrimer/2-hydroxy ethyl methacrylate conjugate in which three regions can be seen. The silane group 48 and 72 protons are in the regions of 0.37 and 0.85 ppm, respectively. The resonance of $CH_2$ protons (c group), $CH_2$, 4 protons (d group), $CH_2$ 8 protons (Heptamer protons), benzylic protons of heptamer ( a group), vinylic $CH_2$ protons are 4.09, 4.15, 4.27, 4.86, 5.66 ppm, respectively.

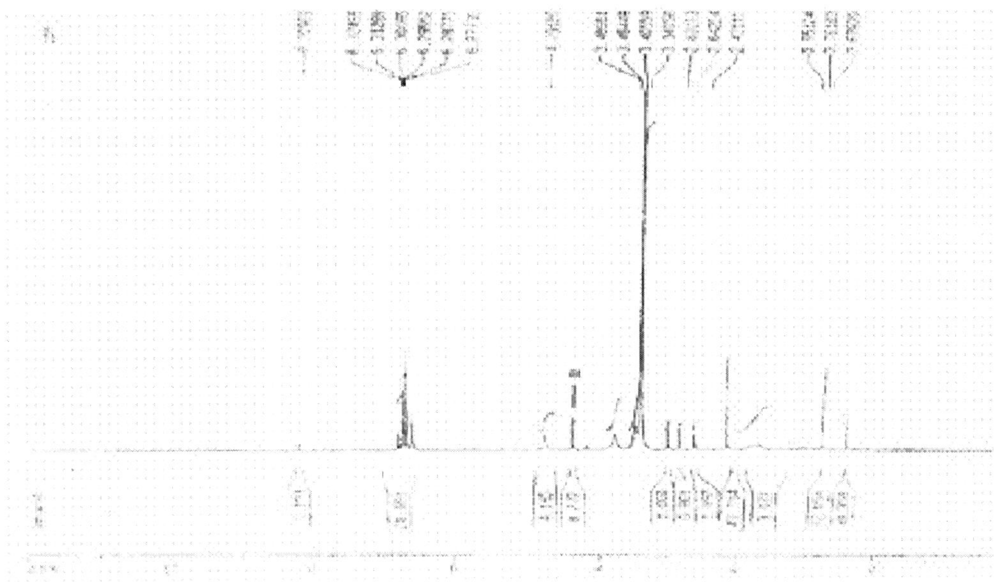

Figure 22. 400MHz 1H-NMR spectrum of [G2] – CH2OH Dendrimer/poly(2-hydroxy ethyl methacrylate ) nanocomposite (DMSO-d6).

The resonances for the aromatic protons of the monomer units on a dendrimer occur in the region 6.73 ppm, separate resonances are observed in the appropriate ratio for each layer of monomer units. The solvent peak occurred within the region of 7.51 ppm.

Figure 22 shows the 400MHz $^{1}$H-NMR spectrum of [G2] – $CH_2OH$ Dendrimer/poly(2-hydroxy ethyl methacrylate)nanocomposite in which three regions can be seen. The silane group 48 and 72 protons are in the regions of 0.10 and 0.85 ppm, respectively. The resonance indicated for the methyl group, polymer $CH_2$ protons, $CH_2O$ protons, $CH_2$ 8 protons (Heptamer protons), $CH_2$ protons of the heptamer (a group), were 1.68, 2.27, 4.17, 4.65, 4.79 ppm, respectively. The resonances for the aromatic protons of the monomer unit in the dendrimer occur in the region 6.28-6.32 ppm, separate resonances are observed in the appropriate ratio for each layer of monomer units. The DMSO peak occurred within the region of 3.34-3.48 ppm.

The thermal behavior of a polymer is important in relation to its properties for controlling the properties of a nanocomposite. The glass transition temperature (Tg) was determined from the DSC thermo grams. The values are given in Table 1. The higher Tg values were probably related to the nanocomposite size, which would decrease the flexibility of the chains and the ability of the chains to undergo segmental motion and would increase the Tg values. On the other hand, the introduction of a strongly phenolic OH group can increase the Tg value because of the formation of internal hydrogen bonds between the polymer chains. Nano and micro polymer bonded drugs (50 mg) were poured into 3 mL of aqueous buffer solution (SGF: pH 1 or SIF: pH 7.4). The mixture was introduced into a cellophane membrane dialysis bag. The bag was closed and transferred to a flask containing 20 mL of the same solution maintained at 37° C. The external solution was continuously stirred, and 3 mL samples were removed at selected intervals. The removed volume was replaced with SGF or SIF. Triplicate samples were used. The sample of hydrolyzate was analyzed and the quantity of 5-ASA was determined by using a standard calibration curve obtained under the same conditions. Results of drug content and encapsulation efficiency (EE) demonstrated that drug content increased from 9.24±0.81 mg/100mg to 25.70±0.43mg/100mg with the increase in the amount of drug from 5% to 20% wt/wt. No increase in drug content was observed further when the amount of drug increased above 15% wt/wt. This could be due to the limited solubility of drug in solvent and this was endorsed by the presence of drug particles on the surface of the nanocomposite with a drug concentration of 20%. In the case of nanocomposite, nearly 80% to 90% of the drug was released in the initial 8 to 10 hours. This situation is not acceptable for drugs that are required to be released locally in the colon. The effect of drug concentration, nanocomposite concentration and cross-linking time on *in vitro* drug release was observed. *In vitro* drug release after 10 hours was 84.2%±4.1% in the microsized particles and pH =7.4 for 15% of the drug, while it was 89.2%±3.78% for nanosized particles with 20% of the drug. The effect of nanocomposite concentration on the release of drug was found to be meager. Also observed was that the amount of drug released from nanocomposite decreased when the cross-linking time increased. These properties are probably explained by the promotion of cross-links between the nanocomposite chains. Freeze-drying the samples resulted in larger and more porous nanocomposite and slightly faster disintegration times compared with air-dried nanocomposite. Freeze-drying had the advantage of avoiding drug extraction by immediately freezing and removing the water present within the nanocomposite by sublimation. When air-dried, the nanocomposite was exposed to water within for longer periods of time. In the second part of this investigation, cross-linked nanocomposite was as nanosized particles. The relatively higher drug release after 10 hours reveals a signification increase. This increase in drug release suggests that most of the drug had dissolved and leached out through in nanocomposite. The drug released during 10-hour release rate studies was because of the presence of nanosized particles. Modified glycol chitosan nanoparticles for cancer therapy have been recently formulated for the administration of docetaxel camptothecin. Graft copolymerization of vinyl monomers onto polysaccharides using free radical initiation, has attracted the interest of many scientists. Up to now, considerable works have been devoted to the grafting of vinyl monomers onto the substrates, especially starch and cellulose. Existence of polar functionally groups as carboxylic acid need not only be for bioadhesive properties but also for polymers with pH-sensitive properties. Because the increase of methacrylic acid content in the hydrogels provides more hydrogen bonds at low pH and more electrostatic repulsion at high pH. Also, chitosan is a functional linear polymer derived from chitin, the most abundant natural polysaccharide on the earth after cellulose, and it is not digested in the upper GI tract by human digestive enzymes. Chitosan is a copolymer consisting of 2-amino-2-deoxy-D-glucose and 2-acetamido-2-deoxy-D-glucose units links with b-(1→ 4) bonds. It should be susceptible to glycosidic hydrolysis by microbial enzymes in the colon because it possesses glycosidic linkages similar to those of other enzymatically depolymerized polysaccharides. Among diverse approaches that are possible for modifying polysaccharides, grafting of synthetic polymer is a convenient method for adding new properties to a polysaccharide with minimum loss of its initial properties. As stated earlier, this is part of our research program on CS/CMS modification to prepare materials with pH-sensitive properties for uses as colon-specific drug delivery. The free radical graft copolymerization of polymethacrylic acid onto CS/CMS was carried out in DMF/$H_2O$ solvent system at 40 °C. A 5-aminosalicylic acid (5-ASA) is an active ingredient of agents used for the long-term maintenance therapy to prevent relapses of Crohn's disease and ulcerative colitis. However, when 5-ASA is administered orally, a large amount of the drug is absorbed in the upper gastrointestinal tract (GIT), and causes systemic side-effects. Therefore, it is preferable to deliver the drug site-specifically to the colon. Research on the colonic drug delivery system can be performed in the stomach, small intestine, cecum, and colon areas. In this section, we have used a CS/CMS/PMA–5-ASA conjugate as the only prodrug for the colon-specific area drug delivery system. The mixture modified hydrogel and 5-aminosalicylic acid (5-ASA) and salicylic acid (SA) as the model drugs that were converted to nano by using the freeze-drying method. The equilibrium swelling studies and *in vitro* release profiles were carried out in enzyme-free simulated gastric and intestinal fluids (SGF and SIF, respectively).The influences of different factors, such as content of methacrylic acid in the feed monomer and swelling were studied.

To achieve successful colonic delivery, a drug needs to be protected from the absorption environment of the upper gastrointestinal tract (GIT). Then, this drug must be abruptly released into the proximal colon, which is considered the optimum site for colon-targeted delivery of drugs. These requirements have prompted the development of polymeric systems that swell minimally under acidic conditions but extensively in basic intestinal medium. The effect of drug concentration, CS/CMS concentration and cross-linking time on *in vitro* drug release was observed. The drug release after 10 hours was 84.2%±4.1% in the microsized particles and pH =7.4 for 15% of the drug, while it was 89.2%±3.78% for nanosized particles with 20% of the drug. The effect of CS/CMS concentration on the drug release was found to be meager. Also observed was that the amount of drug released from CS/CMS decreased when the cross-linking time increased. These properties are probably explained by the promotion of cross-links between CS/CMS chains. Freeze-drying the samples resulted in larger and more porous hydrogel and slightly faster disintegration times compared with air-dried hydrogel. Freeze-drying had the advantage of avoiding drug extraction by immediately freezing and removing the water present within the hydrogel by sublimation. When air-dried, the hydrogel was exposed to water within for longer periods of time. In the second part of this investigation cross-linked CS/CMS was used as nanosized particles. The relatively higher drug release after 10 hours revealed a significant increase. This increase in drug release suggests that most of the drug had dissolved and leached out with the hydrogel. The drug released during the 10-hour release rate studies was because of the presence of nanosized particles. A549 lung epithelial cells were treated with pure FITC labeled hydrogel, pure 5-ASA / SA and FITC-labeled hydrogel-5-ASA or SA complex. The UV-Vis analysis using 5-ASA or SA and FITC calibration curves, of the supernatant and the cell lysate for pure hydrogel suggests that most of hydrogel (~ 90%) enters the lung epithelial carcinoma cells within 1 h. This is seen by a decreasing FITC content in the supernatant and a simultaneous increase in the FITC content in the cell lysate. The hydrogel drug complex showed that the cellular entry profile of the complex hydrogel was comparable to the pure hydrogel, but the drug entry was significantly sped up. In the complex state, 5-ASA appears to enter the cells at rates comparable to the pure hydrogel and significantly faster than the free drug (more than 80% of the 5-ASA or SA complex gets inside the cell in an 1h, whereas it takes about 3h for 80% of pure 5-ASA or SA complex to get inside the cell). The ability of the hydrogel to form a complex with drugs depends on the nature of the hydrogel surface, electrostatic interactions between the hydrogel and the drug, plus the ability of the drug to form a conjugate with the hydrogel through chemical bonding. Therefore, it is possible to manipulate the incorporation process for a given drug by appropriate selection of the hydrogel. One might expect that the

5-aminosalicylic acid or salicylic acid with carboxylic group may form a complex with surface $NH_2$ groups of hydrogel. This was investigated using FT-IR and *in vitro* release studies. At fixed intervals, the supernatant was removed, and fresh RPMI medium is added to the cells, and the cells are shaken for more than 1 min. The cells are subsequently trypsinized and the cell lysate was prepared. This was done to ensure the complete removal of hydrogel and /or drug present (if any) on cell surface. Furthermore, a mass balance on 5- ASA / SA was done based on the initial concentration, and the supernatant concentration as a function of time, and most of the drug was accounted for. Therefore, it is unlikely that appreciable amount of drug was just on the cell surface. In order to study potential application of PBDs containing 5-aminosalicylic acid and SA as pharmaceutically active compounds, we have studied the drug release behavior of the polymers under physiological conditions. The important parameter for the increase in diffusion coefficient is the decrease of the particle size. It appears that the degree of drug release polymers depends on their particle size. On the other hand, the chemical structure of the drug is also an important factor in hydrolytic behavior of polymeric prodrugs. The high different hydrolysis rate for SA compared to 5-ASA at pH 1 and 7.4 can be related to the functional groups along the drug. Also, 5-ASA contains both amine (basic) and carboxylic acid (acidic) functional groups. This factor ultimately results in an increase of hydrophilicity for 5-ASA in pH 1 and 7.4, and reduces a different hydrolysis rate for 5- ASA when compared to SA at pH 1 and 7.4.

The higher Tg values probably related to the introduction of cross-links, which would decrease the flexibility of the chains and the ability of the chains to undergo segmental motion, which would increase the Tg values. On the other hand, the introduction of a strongly polar carboxylic acid group can increase the Tg value because of the formation of internal hydrogen bonds between the polymer chains.

This results in a network of polymers that swell and become soft in solvents such as $H_2O$ and most organic solvents without dissolving. To measure the swelling, preweighted dry drug-free hydrogels were immersed in various buffer solutions (pH 7.4 and pH 1) at 37° C. After excess water on the surface was removed with the filter paper, the weight of the swollen samples was measured at various time intervals.

## Drug Release Based on the Presence of Colonic Microflora

Both anaerobic and aerobic micro-organisms inhabit the human gastrointestinal tract. In the small intestine the microflora is mainly aerobic, but in the large intestine it is anaerobic. About 400 bacterial species have been found in the colon, and some fungi. Most bacteria inhabit in the proximal areas of the large intestine, where energy sources are greatest. Carbohydrates arriving from the small intestine form the main source of nourishment for bacteria in the colon. The carbohydrates are split into short-chain fatty acids, carbon dioxide and other products by the enzymes glycosidase and polysaccharidase. Protease activity in the colon can result in cleavage of proteins and peptides. In the proximal colon the pH is lower than at the end of the small bowel because of the presence of short-chain fatty acids and other fermentation products. Diet can affect colonic pH.

The presence of colonic microflora has formed a basis for the development of colon-specific drug delivery systems. Interest has focused primarily on azo reduction and hydrolysis of glycoside bonds. However, the colonic microflora varies substantially between and within individuals, reflecting diet, age, and disease. Such variations need to be taken into account in developing colon-specific formulations depending on the presence of colonic microflora. There is also significant proteolytic activity in the colon, although this is 20 to 60 times less than in the small bowel. Even when proteolytic activity is relatively low a drug may remain much longer in the colon than in the small intestine, with the result that has it is exposed longer to proteolytic activity.

Prodrugs have been used as targeting drugs for the large intestine. Sulphasalazine, used in the treatment of ulcerative colitis and Crohn's disease, is a colon-specific prodrug. In the colon, sulphasalazine is split by bacterial azoreduction into 5-ASA and sulphapyridine. Sulphapyridine can cause side-effects, and other carriers for delivery of 5-ASA to the colon have therefore also been investigated. Olsalatzine consists of two molecules of 5-ASA linked by an azo-bond. Ipsalatsine and balsalatsine are other 5- ASA containing prodrugs. Polymers and polyamides containing azo groups have been used to convey 5-ASA to the large intestine. Azo polymers have been used as colon specific film coatings. Colon targeting by means of azo polymers is associated with many problems. Microbial degradation of azo polymers is usually slow, and drug delivery can be incomplete and irregular. Not enough is yet known about the safety of azo polymers. *In vivo* absorption studies with azo polymers have mostly been carried out using rats. No results are available for human beings using this method. Although the gastrointestinal microflora of rats and humans differ, results of *in vivo* experiments with rats can give some indications regarding biodegradation of azo polymers.

Hydrogels containing azo-aromatic cross-links have been investigated in connection with site-specific drug delivery of peptide and protein drugs. In the low pH range of the stomach the gels have a low equilibrium degree of swelling and the drug is protected against digestion by enzymes, but at high pH levels they swell. So in the stomach a drug will be protected, but released in the colon, where cross-links become degraded. The colonic microflora produces a wide range of glycosidases capable of hydrolyzing glycosides and polysaccharides. Glycosides of glucocorticosteroids have been synthesized, and tested in rodents. The problem in these studies was that some drug was hydrolyzed even in the small intestine. However, in rodent bacterial glycosidase activity in the small intestine is some 100 times greater than in human beings. It is likely that drug delivery for a human would be more predictable than in rodents. Glucuronides, which are less subject to hydrolysis in the small intestine than glycosides, have also been used as drug carriers. An extensive range of drug delivery systems based on polysaccharides has been investigated. The advantage of these materials is that most are easily available. Disadvantages are that most of polysaccharides are hydrophilic and gel forming. In preparing dosage forms from polysaccharides it is necessary to ensure that no drug is released until it reaches the colon.

Amylose has been used in coatings of colon-specific formulations. Amylose, a major component of starch, swells too much on its own, but amylose-ethylcellulose coatings have been investigated in connection with targeting a drug release on the colon. From the results of *in vitro* studies it was concluded that amylose-ethylcellulose coatings could be suitable for colon-specific formulations. Pectin is a polysaccharide, found in the cell walls of plants. It is totally degraded by colonic bacteria but is not digested in the upper gastrointestinal tract. One disadvantage of pectin is its solubility. This can however be adjusted by changing its degree of methoxylation, or by preparing calcium pectinate. The film-coating properties of pectin have been improved through use of ethylcellulose. Pectin has also been used with chitosan and HMPC. It has been shown in studies in which gamma camera was used that determined pectin-coated tablets disintegrate in the colon during transit. Cross-linked guar gum has been used as a drug carrier in matrix tablets. It was concluded that guar gum is suitable for preparation of colon-specific formulations and is particularly suitable as a carrier of drugs that are not very soluble in water. However, the guar gum formulations mentioned have only formed in subjects for *in vitro* dissolution studies and *in vivo* evaluation in rats. Dextran ester prodrugs have been investigated as means of transporting drugs to the colon. When the bioavailability of naproxen after administration of dextran-naproxen prodrug was assessed in pigs, lag times of two to three hours were observed.

Dextran esters of fatty acids have been used to form colon-specific film coatings. The suitability of such formulations for colon-specific drug delivery on human beings, still remain to be demonstrated by volunteers. Chitosan is a high-molecular-weight polysaccharide that is degraded by colonic microflora. Insulin and 5-ASA have been administered to rats in enteric-coated chitosan capsules. A multiple-unit formulation containing chitosan and drug has also been prepared. This formulation depended for drug delivery on both variations in gastrointestinal pH and the presence of colonic microflora.

## Pressure-Controlled Drug-Delivery Systems

As a result of peristalsis, higher pressures are encountered in the colon than in the small intestine. Takaya et al., developed pressure controlled colon-delivery capsules prepared using an ethylcellulose, which is insoluble in water. In such systems drug release occurs following disintegration of a water-insoluble polymer capsule as a result of pressure in the lumen of the colon.

The thickness of the ethyl cellulose membrane is the most important factor for disintegration of the formulation. The system also appeared to depend on capsule size. When salivary secretion of caffeine after oral administration of pressure-controlled capsules was studied in human volunteers, a correlation was found between ethyl cellulose membrane thickness and the time interval in the appearance of caffeine in the saliva.

Because of reabsorption of water from the colon, the viscosity of luminal content is higher in the colon than in the small intestine. It has therefore been concluded that drug dissolution in the colon could present a problem in relation to colon-specific oral drug delivery systems. In pressure-controlled ethyl cellulose single-unit capsules the drug is a liquid. Lag times of three to five hours in relation to drug absorption were noted when pressure-controlled capsules were administered to human subjects. It was concluded that the capsules disintegrated in the colon because of an increase in pressure. It was also concluded that the formulation studied was advantageous since the drug release mechanism is independent of pH.

The site at which the formulations disintegrated was not demonstrated in the studies previously mentioned. The mechanism of disintegration was also not clarified. As discussed earlier, ethyl cellulose coatings have also been used in connection with time controlled drug delivery. Disintegration of the formulation can therefore also occur sometime after administration, even in the small intestine.

## Enteric Delivery Systems

The enteric polymers used were the methacrylate copolymers Eudragit™ S and Eudragit™ L (Röhm Pharma, Germany), the hydroxypropyl methylcellulose acetate succinates, Aqoat™ AS-HF and Aqoat™ AS-MF (Shin-Etsu Chemical Co., Japan) and cellulose acetate phthalate Aquateric™ (FMC Corporation, USA). The enteric polymers used dissolve around or below pH 7 (Peeters and Kinget, 1993; Bauer and Kesselhut, 1995), cellulose acetate phthalate at pH 6.2, Eudragit™ L at pH 6 and Eudragit™ S and Aqoat™ AS-HF at pH 7.

## REFERENCES

[1] Adkin, D. A., Davis, S. S., Sparrow, R. A. & Wilding, I. R. (1993). Colonic transit of different sized tablets in healthy subjects. *J. Controlled Rel., 23*, 147-156.

[2] Adkin, D. A., Kenyon, C. J., Lerner, E. I., Landau, I., Strauss, E., Caron, D., Penhasi, A., Rubinstein, A. & Wilding, I. R. 1997. The use of scintigraphy to provide "proof of concept" for novel polysaccharide preperations designed for colonic drug delivery. *Pharm. Res., 14*, 103-107.

[3] Ashford, M. & Fell, J. T. (1993). Colonic delivery of drugs. Current status on targeted drug delivery to the gastrointestinal tract. Short Hills (NJ). Capsugel, Symposia Series, 133-142.

[4] Ashford, M., Fell, J. T., Attwood, D., Sharma, H. & Woodhead, P. J. (1993a). An *in vivo* investigation into the suitability of pH-dependent polymers for colonic targeting. *Int. J. Pharm., 95*, 193-199.

[5] Ashford, M., Fell, J. T., Attwood, D. & Woodhead, P. J. (1993b). An *in vitro* investigation into the suitability of pH-dependent polymers for colonic targeting. *Int. J. Pharm., 91*, 241-245.

[6] Ashford, M., Fell, J. T., Attwood, D., Sharma, H. & Woodhead, P. J. (1994). Studies on pectin formulations for colonic drug delivery. *J. Conrolled Rel., 30*, 225-232.

[7] Avgerinos, A. & Hutt, A. (1986). High-performance liquid chromatographic determination of ibuprofen in human plasma and urine by direct injection. *J. Chromatogr., 380*, 468-471.

[8] Bauer, K. H. & Kesselhut, J. F. (1995). Novel pharmaceutical excipients for colon targeting. *S.T.P. Pharma Sci., 5*, 54-59.

[9] Beerman, B. (1982). Kinetics and dynamics of furosemide and slow-acting furosemide. *Clin. Pharmacol. Ther., 32*, 584-591.

[10] Brønstedt, H. & KopeMek, J. (1992). Hydrogels for site-specific drug delivery to the colon: *In vitro* and *in vivo* degradation. *Pharm. Res., 9*, 1540-1545.

[11] Brockmeier, D., Voegele, D. & von Hattingberg, H. M. (1983). *In vitro-in vivo* correlation, a time scaling problem? *Arzneim. Forsch./Drug Res., 33*, 598-601.

[12] Bussemer, T., Otto, I. & Bodmeier, R. (2001). Pulsatile drug-delivery systems. *Crit. Rev. Ther. Drug Carrier Syst., 18*, 433-458.

[13] Cole, E., Scott, R., Connor, A., Wilding, I., Petereit, H-U., Schminke, C., Beckert, T. & Cadé, D. (2002). Enteric coated HPMC capsules designed to achieve intestinal targeting. *Int. J. Pharm.*, *231*, 83-95.

[14] Davis, S. S., Hardy, J. G. & Fara, J. W. (1986). Transit of pharmaceutical dosage forms through the small intestine. *Gut*, *27*, 886-892.

[15] Dew, M. J., Hughes, P. J., Lee, M. G., Evans, B. K. & Rhodes, J. (1982). An oral preparation to release drugs in the human colon. *Brit. J. Clin. Pharm.*, *14*, 405-408.

[16] Evans, D. F., Pye, G., Bramley, R., Clark, A. G., Dyson, T. J. & Hardcastle, J. D. (1988). Measurement of gastrointestinal pH profiles in normal ambulant human subjects. *Gut, 29*, 1035-1041.

[17] Wei, S., Yan, W., Chen, D. & Li, J. (2001). An investigation of pulsatile release tablets with ethylcellulose and Eudragit L as film coating materials and cross-linked polyvinylpyrrolidone in the core tablets. *J. Controlled Rel.*, *77*, 245-251.

[18] Fallingborg, J., Christensen, L.A., Ingeman-Nielsen, M., Jacobsen, B.A., Abildgaard, K. & Rasmussen, H.H. (1989). pH-profile and regional transit times of the normal gut measured by a radiotelemetry device. *Aliment. Pharmacol. Ther.*, *3*, 605-613.

[19] Friend, D. R. (1991). Colon-specific drug delivery. *Adv. Drug Deliv. Rev.*, *7*, 149-199.

[20] Friend, D. R. (1995). Glycoside prodrugs: Novel pharmacotherapy for colonic diseases. *S.T.P. Pharma Sci.*, *5*, 70-76.

[21] Friend, D. R. & Tozer, T. N. (1992). Drug glycosides in oral colon-specific drug delivery. *J. Controlled Rel.*, *19*, 109-120.

[22] Fukui, E., Miyamura, N. & Kobayashi, M. (2001). An *in vitro* investigation of the suitability of presscoated tablets with hydroxypropylmethylcellulose acetate succinate (HPMCAS) and hydrophobic additives in the outer shell for colon targeting. *J. Controlled Rel.*, *70*, 97-107.

[23] Fukui, E., Miyamura, N., Uemura, K. & Kobayashi, M. (2000). Preparation of enteric-coated timed release press-coated tablets and evaluation of their function by *in vitro* and *in vivo* tests for colon targeting. *Int. J. Pharm.*, *204*, 7-15.

[24] Gabr, K. E. (1992). Effect of organic acids on the release patterns of weakly basic drugs from inert sustained release matrix tablets. *Eur. J. Pharm. Biopharm.*, *38*, 199-202.

[25] Gazzaniga, A., Iamartino, P., Maffione, G. and Sangalli, M. E. (1994a). Oral delayed-release system for colonic specific delivery. *Int. J. Pharm.*, *108*, 77-83.

[26] Gazzaniga, A., Sangalli, M. E. & Giordano, F. (1994b). Oral chronotopic drug delivery systems: Achievement of time and/or site specificity. *Eur. J. Pharm. Biopharm.*, *40*, 246-250.

[27] Gliko-Kabir, I., Yagen, B., Penhasi, A. & Rubinstein, A. (1998). Low swelling, cross-linked guar and its potential use as colon-specific drug carrier. *Pharm. Res.*, *15*, 1019-1025.

[28] Gupta, V., Beckert, T. & Price, J. (2001). A novel pH- and time-based multi-unit potential colonic drug delivery system. I. Development. *Int. J. Pharm.*, *213*, 83-91.

[29] Halsas, M., Hietala, J., Veski, P., Jürjenson, H. & Marvola, M. (1998). Morning vs. evening dosing of ibuprofen in two different oral drug formulations. *Proceedings of 2nd World Meeting on Pharmaceutics, Biopharmaceutics and Pharmaceutical Technology, Paris* 25.-28.5. , 1197-1198.

[30] Harboe, E., Larsen, C., Johansen, M. & Olesen, P. (1989). Macromolecular prodrugs. XIV. Absorption characteristics of naproxen after oral administration of a dextran T-70-naproxen ester prodrug in pigs. *Int. J. Pharm.*, *53*, 157-165.

[31] Hardy, J. G., Healey, J. N. C. & Reynolds, J. R. (1987). Evaluation of an enteric-coated delayed-release 5-aminosalicylic acid tablet in patients with inflammatory bowel disease. *Aliment. Pharmacol. Ther.*, *1*, 273-280.

[32] Hata, T., Shimazaki, Y., Kagayama, A., Tamura, S. & Ueda, S. (1994). Development of a novel drug delivery system, time-controlled explosion system (TES). Part 5. Animal pharmacodynamic and human bioavailability study. *Int. J. Pharm.*, *110*, 1-7.

[33] Hebden, J. M., Perkins, A. C. & Spiller, R. C. (1999). Scintigraphic study of colonic release and absorption. In: A.C. Perkins and M. Frier (eds.), *Nuclear Medicine in Pharmaceutical Research*. Taylor and Francis. London, 101-112.

[34] Herzfeldt, C. D. & Kümmel, R. (1983). Dissociation constants, solubilities and dissolution rates of some selected nonsteroidal anti-inflammatories. *Drug Dev. Ind. Pharm.*, *9*, 767-793.

[35] Hosny, E., Al-Shora, H. & Elmazar, M. (2002). Oral delivery of insulin from enteric-coated capsules containing salicylate: Effect on relative hypoglycemia of diabetic beagle dogs. *Int. J. Pharm.*, *237*, 71-76.

[36] Hu, Z., Kimura, G., Ito, Y., Mawatari, S., Shimokawa, T., Yoshikawa, H., Uoshikawa, Y. & Takada, K. (1999). Technology to obtain sustained release characteristics of drugs after delivered to the colon. *J. Drug Target*, *6*, 439-448.

[37] Hu, Z., Mawatari, S., Shimokawa, T., Kimura, G., Yoshikawa, Y., Shibata, N. & Takada, K. (2000). Colon delivery efficiencies of intestinal pressure-contolled colon delivery capsules prepared by a coating machine in human subjects. *J. Pharm. Pharmacol.*, *52*, 1187-1193.

[38] Hunter, E., Fell, J. T. & Sharma, H. (1982). The gastric emptying of pellets contained in hard gelatin capsules. *Drug Dev. Ind. Pharm.*, *8*, 751-757.

[39] Hunter, E., Fell, J. T. & Sharma, H. (1983). The gastric emptying of hard gelatin capsules. *Int. J. Pharm.*, *17*, 59-64.

[40] Ishibashi, T., Hatano, H., Kobayashi, M., Mizone, M. & Yoshino, M. (1998a). Design and evaluation of a new capsule-type dosage form for colon-targeted delivery of drugs. *Int. J. Pharm.*, *168*, 31-40.

[41] Ishibashi, T., Pitcairn, G., Yoshino, H., Mizobe, M. & Wilding, I. (1998b). Scintigraphic evaluation of a new capsule-type colon specific drug delivery system in healthy volunteers. *J. Pharm. Sci.*, *87*, 531-535.

[42] Jeong, Y-I., Ohno, T., Hu, Z., Yoshikawa, Y., Shibata, N., Nagata, S. & Takada, K. (2001). Evaluation of an intestinal pressure-controlled colon delivery capsules prepared by a dipping method. *J. Controlled Rel.*, *71*, 175-182.

[43] Khan, M., Prebeg, Ž. & Kurjakovi, M. N. (1999). A pH-dependent colon targeted oral drug delivery system using methacrylic acid copolymers. I. Manipulation of drug release using Eudragit™ L100- 55 and Eudragit™ S100 combinations. *J. Controlled Rel.*, *58*, 215-222.

[44] Kinget, R., Kalala, W., Vervoort, L. & van den Mooter, G. (1998). Colonic drug delivery. *J. Drug Target*, *6*, 129-149.

[45] Klotz, U. (1985). Clinical pharmacokinetics of sulphasalazine, its metabolites and other prodrugs of 5-aminosalisylic acid. *Clin. Pharmacokinet.*, 10, 285-302.

[46] KopeMek, J., KopeMeková, P., Brøndstedt, H., Rathi, R., Dihová, B., Yeh, P.-Y. & Ikesue, K. (1992). Polymers for colon-specific drug delivery. *J. Controlled Rel.*, *19*, 121-130.

[47] Krögel, I. & Bodmeier, R. (1998). Pulsatile drug release from an insoluble capsule body controlled by an erodible plug. *Pharm. Res.*, *15*, 474-481.

[48] Leopold, C. & Eikeler, D. (2000). Basic coating polymers for the colon-specific drug delivery in inflammatory bowel disease. *Drug Dev. Ind. Pharm.*, *26*, 1239-1246.

[49] Lorenzo-Lamosa, M., Remuñán-López, C., Vila-Jato, J. & Alonso, M. (1998). Design of microencapsulated chitosan microspheres for colonic drug delivery. *J. Controlled Rel.*, *52*, 109-118.

[50] Luckow, V. (1994). *In vitro/in vivo* correlation in custom tailoring of sustained release drug formulations. *Bulletin Technique Gattefosse*, *86*, Saint Priest (France), 33-41.

[51] McLeod, A., Friend, D. & Tozer, T. (1993). Synthesis and chemical stability of glucocorticoid-dextran esters: Potential prodrugs for colon-specific delivery. *Int. J. Pharm.*, *92*, 105-114.

[52] Milojevic, S., Newton, J., Gummings, J., Gibson, G., Botnam, L., Ring, S., Stockham, M. & Allwood, C. (1996a). Amylose as a coating for drug delivery to the colon: Preparation and *in vitro* evaluation using 5-aminosalicylic acid pellets. *J. Controlled Rel.*, *38*, 75-84.

[53] Milojevic, S., Newton, J., Gummings, J., Gibson, G., Botnam, L., Ring, S., Stockham, M. & Allwood, C. (1996b). Amylose as a coating for drug delivery to the colon: Preparation and *in vitro* evaluation using glucose pellets. *J. Controlled Rel.*, *38*, 85-94.

[54] Munjeri, O., Collett, J. H. & Fell, J. T. (1997). Hydrogel beads based on amidated pectins for colon specific drug delivery: The role of chitosan in modifying drug release. *J. Controlled Rel.*, *46*, 273-278.

[55] Muraoka, M., Hu, Z., Shimokawa, T., Sekino, S., Kurogoshi, R., Kuboi, Y., Yoshikawa, Y. & Takada, K. (1998). Evaluation of intestinal pressure-controlled colon delivery capsule containing caffeine as a model drug in human volunteers. *J. Controlled Rel.*, *52*, 119-129.

[56] Narisawa, S., Nagata, M., Danyoshi, C., Yoshino, H., Murata, K., Hirakawa, Y. & Noda, K. (1994). An organic acid-induced sigmoidal release system for oral controlled-release preparations. *Pharm. Res.*, *11*, 111-116.

[57] Narisawa, S., Nagata, M., Hirakawa, Y., Kobayashi, M. & Yoshino, H. (1996). An organic acid-induced sigmoidal release system for oral controlled-release preparations. 2. Permeability enhancement of Eudragit™ RS coating led by the physicochemical interactions with organic acid. *J. Pharm. Sci.*, *85*, 184-188.

[58] Narisawa, S., Nagata, M., Hirakawa, Y., Kobayashi, M. & Yoshino, H. (1997). An organic acid-induced sigmoidal release system for oral controlled-release preparations. 3. Elucidation of the anomalous drug release behavior through osmotic pumping mechanism. *Int. J. Pharm.*, *148*, 85- 91.

[59] Parker, G., Wilson, C. G. & Hardy, J. G. (1988). The effect of capsule size and density on transit through the proximal colon. *J. Pharm. Pharmacol.*, *40*, 376-377.

[60] Peeters, R. & Kinget, R. (1993). Film-forming polymers for colonic drug delivery: I. Synthesis and physical and chemical properties of methyl derivatives of Eudragit™ S. *Int. J. Pharm.*, *94*, 125-134.

[61] Pozzi, F., Furlani, P., Gazzanica, A., Davis, S. S. & Wilding, I. R. (1994). The time clock system: A new oral dosage form for fast and complete release of drug after a predetermined lag time. *J. Controlled Rel.*, *31*, 99-108.

[62] Rama Prasad, Y. V., Krishnaiah, Y. S. R. & Satyanarayana, S. (1998). *In vitro* evaluation of guar gum as a carrier for colon-specific drug delivery. *J. Controlled Release*, *51*, 281-287.

[63] Rao, S. S. & Ritschel, W. A. (1992). Development and *in vitro / in vivo* evaluation of a colonic release capsule of vasopressin. *Int. J. Pharm.*, *86,* 35-41.

[64] Rittschell, W. A. (1992). *Handbook of Basic Pharmacokinetics*, 4th edit., Drug Intelligence Publications, Hamilton Press, Illinois.

[65] Ritschell, W. A., Menon, A. & Sakr, A. (1991). Biopharmaceutical evaluation of furosemide as a potential candidate for a modified release peroral dosage form. *Meth. Find Exp. Clin. Pharmacol.*, *13*, 629-636.

[66] Ross, A., Macrae, R., Walther, M. & Stevens, H. (2000). Chronopharmaceutical drug delivery from a pulsatile capsule device based on programmable erosion. *J. Pharm. Pharmacol.*, *52*, 903-909.

[67] Rubinstein, A. (1990). Microbially controlled drug delivery to the colon. *Biopharm. Drug Dispos.*, *11*, 465-475.

[68] Rubinstein, A. (1995). Approaches and opportunities in colon-specific drug delivery. *Crit. Rev. Ther. Drug Carrier Syst.*, *12*, 101–149.

[69] Rubinstein, A., Radai, R., Ezra, M., Pathak, S. & Rokem, S. (1993). *In vitro* evaluation of calcium pectinate: A potential colon-specific drug delivery carrier. *Pharm. Res.*, *10*, 258-263.

[70] Rudolph, M., Klein, S., Beckert, T., Petereit, H-U. & Dressman, J. (2001). A new 5-aminosalicylic acid multi-unit dosage form for the therapy of ulcerative colitis. *Eur. J. Pharm. Biopharm.*, *51*, 183-190.

[71] Saffran, M., Sudesh Kumar, G., Savariar, C., Burnham, J., Williams, F. & Neckers, D. (1986). A new approach to the oral administration of insulin and other peptide drugs. *Science*, *233*, 1081-1084.

[72] Sangalli, M.E., Maroni, A., Zema, L., Busetti, C., Giordano, F. & Gazzanica, A. (2001). *In vitro* and *in vivo* evaluation of an oral system for time and/or site-specific drug delivery. *J. Controlled Rel.*, *73*, 103-110.

[73] Schacht, E., Gevaert, A., Kenawy, E. R., Molly, K., Verstraete, W., Adriaensens, P., Carleer, R. & Gelan, J. (1996). Polymers for colon specific drug delivery. *J. Controlled Rel.*, *39*, 327-338.

[74] Spitael, J. & Kinget, R. (1979). Solubility and dissolution rate of enteric polymers. *Acta Pharm. Technol.*, *7*, 163-168.

[75] Stevens, H., Wilson, C., Welling, P., Bakhshaee, M., Binns, J., Perkins, A., Frier, M., Blackshaw, E., Frame, M., Nichols, D., Humphrey, M. & Wicks, S. (2002). Evaluation of Pulsincap™ to provide regional delivery of dofetilide to the human GI tract. *Int. J. Pharm.*, *236*, 27-34.

[76] Takaya, T., Ikeda, C., Imagawa, N., Niwa, K. & Takada, K. (1995). Development of a colon delivery capsule and the pharmacological activity of recombinant human granulocyte colony-stimulating factor (rhG-CSF) in beagle dogs. *J. Pharm. Pharmacol.*, *47*, 474-478.

[77] Takaya, T., Niwa, K., Muraoka, M., Ogita, I., Nagai, N., Yano, R., Kimura, G., Yoshikawa, Y., Yoshikawa, H. & Takada, K. (1998). Importance of dissolution process on systemic availability of drugs delivered by colon delivery system. *J. Controlled Rel.*, *50*, 111-122.

[78] Touitou, E. & Rubinstein, A. (1986). Targeted enteral delivery of insulin to rats. *Int. J. Pharm.*, *30*, 95-99.

[79] Tozaki, H., Komoike, J., Tada, C., Maruyama, T., Terabe, A., Suzuki, T., Yamamoto, A. & Muranishi, S. (1997). Chitosan capsules for colon-specific drug delivery: Improvement of insulin absorption from the rat colon. *J. Pharm. Sci.*, *86*, 1016-1021.

[80] Turkoglu, M., Takka, S., Baran, H. & Sakr, A. (1999). Pectin-hydroxypropyl-methylcellulose drug delivery system for colon targeting. *Pharm. Ind.*, *61*, 662-665.

[81] USP 23, The United States Pharmacopeia, (1994). United States Pharmacopeial Convention, Inc., Rockville, MD. Wakerly, Z., Fell, J. T., Attwood, D. & Parkins, D. (1996). Pectin/Ethylcellulose film coating formulations for colonic drug delivery. *Pharm. Res.*, *13*, 1210-1212.

[82] Vandelli, M. A., Leo, E., Forni, F. & Bernabei, M. T. (1996). *In vitro* evaluation of a potential colonic delivery system that releases drug after a controllable lag-time. *Eur. J. Pharm. Biopharm.*, *43*, 148-151.

[83] Van den Mooter, G., Maris, B., Samyn, C., Augustijns, P. & Kinget, R. (1997). Use of azo polymers for colon-specific drug delivery. *J. Pharm. Sci.*, *86*, 1321-1327.

[84] Van den Mooter, G., Samyn, C. & Kinget, R. (1992). Azo polymers for colon-specific delivery. *Int. J. Pharm.*, *87*, 37-46.

[85] Washington, N., Washington, C. & Wilson, C. (2001). Physiological Pharmaceutics. Taylor and Francis. London. Watts, P. & Illum, L. (1997). Colonic drug delivery. *Drug Dev. Ind. Pharm.*, *23*, 893-913.

[86] Wilding, I., Coupe, A. & Davis, S. (1991). The role of g-scintigraphy in oral drug delivery. *Adv. Drug Del. Rev.*, *7*, 87-117.

[87] Wilding, I. R., Davis, S. S., Bakhshaee, M., Stevens, H. N. E., Sparrow, R. A. & Brennan, J. (1992). Gastrointestinal transit and systemic absorption of captopril from a pulsed-release formulation. *Pharm. Res.*, *9*, 654-657.

[88] Wilson, C. G., Washington, N., Greaves, J. L., Kamali, F., Rees, J. A., Sempik, A. K. & Lampard, J. F. 1(989a). Bimodal release of ibuprofen in a sustained-release formulation: A scintigraphic and pharmacokinetic open study in healthy volunteers under different conditions of food intake. *Int. J. Pharm.*, *50*, 155-161.

[89] Wilson, C. G., Washington, C. & Washington, N. (1989b). The stomach: Its role in oral drug delivery. In: C. G. Wilson and N. Washington,(eds.), *Physiological Pharmaceutics,* Ellis Horwood Limited, Chichester. 47-70.

[90] Yang, L., Chu, J. & Fix, J. (2002). Colon-specific drug delivery: New approaches and *in vitro/in vivo* evaluation. *Int. J. Pharm.*, *235*, 1-15.

[91] Zahirul, M., Khan, I., Prebeg, Ž. & Kurjakovih, N. (1999). A pH-dependent colon targeted oral drug delivery system using methacrylic acid copolymers. I. Manipulation of drug release using Eudragit™ L100-55 and Eudragit™ S 100 combinations. *J. Controlled Rel.*, *58*, 215-222.

*Chapter 2*

# MUCOSAL DELIVERY SYSTEMS

## NASAL DELIVERY SYSTEMS

Nasal route as a source for delivering drugs to the body is not a recent discovery. In fact, this method for administering drugs has been used for a very long time. Ayurved had written about 5,000 years ago, the reference of the 'Nasya karma' for the administration of drugs. In recent years, a growing interest has focused on the use of the nasal route as a systemic delivery for targeting the brain to administer compounds for local pharmacological effects. In this aspect, an understanding of the deposition, residence time, and clearance of drug formulations from the human nasal cavity is important in order to optimally exploit the nasal route for the delivery of drugs. The nasal cavity possesses many advantages as a site for drug delivery, such as the ease of administration and avoidance of first pass metabolism by the liver through direct absorption into the systemic circulation.

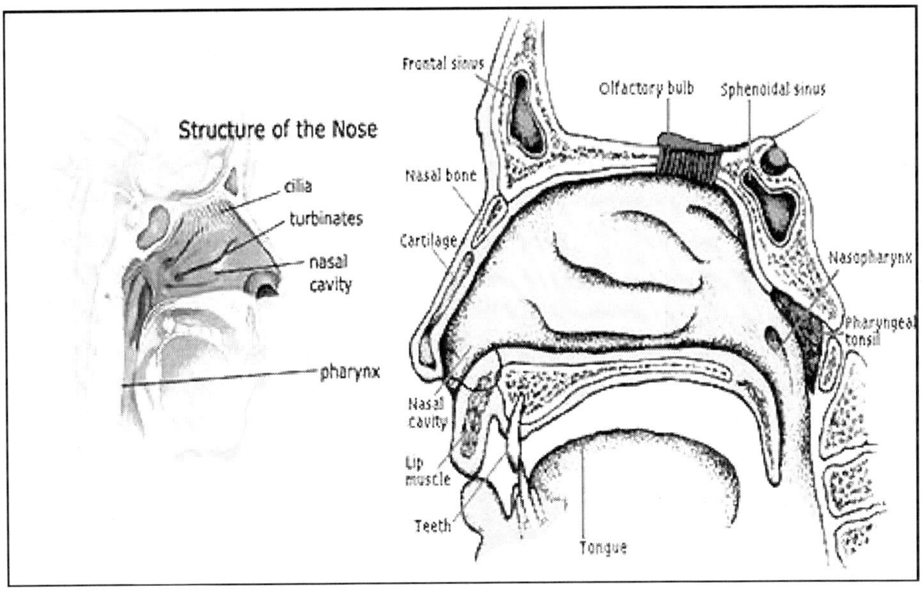


Figure 1. This illustrates the Nose to Brain – Anatomy Physiology.

The nasal drug delivery system comprises of targeting a drug through the nasal epithelium. This mode of drug delivery has gained recent interest due to the tremendous absorptive potential of the nasal mucosa owing to its high permeability because of high perfusion rate. This route can be selectively exploited for delivery of small molecules, peptides and proteins that are not easily administered by routes other than i.v. where immediate response is desired. In the nasal cavity, the respiratory region has the highest degree of vascularity. The olfactory region is located in the top of the nasal cavity and it is the only site of the body where the CNS is in contact with the external environment. The nasal cavity is divided into non-connected parts by a midline septum. Three regions can be distinguished in each part: (1) vestibular region responsible for filtering out the air borne particles, located just inside the nostrils, (2) respiratory region – containing the nasal turbinate and having the highest degree of vascularity, (3) olfactory region – located in the top of the nasal cavity. This is the only physiological site where the brain is in direct contact with surrounding environment and hence exposed to volatile odorants and airborne substances. Administered drugs can reach the brain by traversing one of the two physiological barriers, Blood Brain Barrier (BBB), and Cerebrospinal fluid (CSF). The BBB has many properties similar to a continuous cell membrane, allowing lipid-soluble molecular transport across the membrane where hydrophilic solutes demonstrate minimal permeation. The blood brain barrier (BBB) is one of the strictest barriers of *in vivo* therapeutic drug delivery. The barrier allows restricted exchange of hydrophilic compounds, small proteins, and charged molecules between plasma and central nervous system (CNS).

Lipophillic drugs are transported by the transcellular pathway and cross the BBB to reach the brain. Hydrophilic drugs are transported by paracellular pathway. When polar molecules reach the blood stream, they get a very poor chance of passing the blood brain barrier; otherwise the carrier must mediate the transport mechanism. The polar molecules diffuses slowly from the respiratory region into the blood stream, they have some chances of reaching the olfactory mucosa and from there can transport to CNS. Where a nose-to-brain pathway for a drug like Apomorphine, by intranasal delivery, significantly increases brain accumulation of Apomorphine and could be an effective alternative to subcutaneous and oral formulations. The olfactory epithelium is a gateway for substances entering the CNS and peripheral circulation. The neural connections between the nasal mucosa and the brain provide a unique pathway for non-invasive delivery of therapeutic agents to CNS. The olfactory neural pathway provides both an intraneuronal and extraneuronal pathway into the brain. The intraneuronal pathway involves axonal transport and requires hours to days for drugs to reach different brain regions. Whereby, the extraneuronal pathway probably relies on bulk flow transport through perineural channels, which deliver drugs directly to the brain parenchymal tissue and/or CSF. The extraneuronal pathway allows therapeutic agents to reach the CNS within minutes. Intranasal delivery of agents to the CSF is not surprising as CSF normally drains along the olfactory axon bundles as they traverse the cribriform plate of skull and approaches the olfactory submucosa in the roof of the nasal cavity, where CSF is then diverted into the nasal lymphatics. Thorne et al., reported that trigeminal neural pathway may also be involved in rapidly delivering protein therapeutic agents, such as insulin-like growth factor-1 to the brain and spinal cord following intranasal administration. The transport of drugs across the nasal membrane and into the blood stream may involve either passive diffusion of drug molecules through the pores in the nasal mucosa or some form of non-passive transport.

## DRUG DISTRIBUTION

Drug distribution in the nasal cavity is an important factor that affects the efficiency of nasal absorption. The mode of drug administration may affect this distribution, which in turn can help to determine the extent of absorption of a drug. Nasal deposition of particles is related to the individual's nasal resistance to airflow. For nasal breathing, nearly all the particles having an aerodynamic size of 10-20 mm are deposited on the nasal mucosa. The deposition of particles in the respiratory tract is a function of particle size and respiratory patterns.

The density, shape, and hygroscopicity of particles also affect the deposition. The pathological conditions in the nasal passage will also influence the deposition of particles, whereas the particle-size distribution will determine the site of deposition and affect the subsequent biological response in animals and humans. Furthermore, improvement of the delivery system and drug formulation is necessary to achieve a better clinical effect and easier handling by patients.

Three mechanisms are usually considered in assessing particle deposition in the respiratory tract, *i.e.*, inertia, sedimentation, and diffusion; the first being the dominant mechanism in nasal deposition. Any particle with an aerodynamic diameter of 50 mm or greater does not enter the nasal passage. It was demonstrated that 60% of aerosolized particles of 2-20 mm are deposited in the anterior regions of the nostrils, 2-3 mm from the external areas. The site of drug deposition within the nasal cavity depends on the type of delivery system and the technique used in its application.

## DRUG ABSORPTION

The first step in the absorption of drugs from the nasal cavity is the passage through the mucus. Small, uncharged particles easily pass through this layer. However, larger or charged particles may find it more difficult to cross. Mucin, the principal protein in the mucus, has the potential to bind to solutes and hinder its diffusion. Additionally, structural changes in the mucus layer are possible as a result of environmental changes (*i.e.*, pH and temperature). After a drug's passage through the mucus, there are several mechanisms for absorption through the mucosa. These include transcellular or simple diffusion across the membrane, paracellular transport via movement between cells, and transcytosis by vesicle carriers. Drug absorption obstacles stem from the potentials that occur in the metabolism before reaching the systemic circulation and limited residence time in the cavity. Nasal absorption is affected by molecular weight, size, the pH, the pKa molecule, and delivery volume, among other formulation characteristics. Molecular weight still presents the best correlation to absorption. The apparent cut-off point for molecular weight is approximately 1,000 daltons, with molecules less than 1,000 having better absorption. Shape is also important. Linear molecules have lesser absorption than cyclic-shaped molecules. Additionally, particles should be larger than 10 μm, otherwise the drug may be deposited in the lungs. Hydrophilicity has been found to decrease the drug's bioavailability. The pH is also an important formulation factor for drug absorption.

Both the pH of the nasal cavity and pKa of a particular drug need to be considered to optimize systemic absorption. Nasal irritation is minimized when products are delivered with a pH range of 4.5 to 6.5. Volume and concentration are also important considerations. The delivery volume is limited by the size of the nasal cavity. An upper limit of 25 mg/dose and a volume of 25 to 150 mL/nostril have been suggested.

### Factors Influencing Nasal Absorption

The factors that affect the permeability of a drug through the nasal mucosa can be broadly classified as biological, formulation, and device related.

## BIOLOGICAL FACTORS

The barriers created by cellular arrangement, enzyme milieu, and the physiological status of the mucus membranes greatly influence drug absorption. Some of these factors will be discussed in this section. Nasal absorption shows regional variation because different types of cells constitute the nasal epithelium and because of other factors such as the presence of microvilli, cell density, surface area, and the number of cells.

The respiratory region is richly supplied with blood, has a large surface area and receives the maximum amount of nasal secretions; rendering it most suitable for the permeation of compounds. Ultra structural studies have shown fragmentation and discontinuity in tight junctions around filled goblet cells and this discontinuity could be of relevance to the absorption of aerosol drugs deposited on the airway epithelium. Enzymes such as peptidases and proteases present in the lumen of the nasal cavity or in the epithelial barrier limit the absorption of drugs such as calcitonin, insulin, luteinizing hormone – releasing hormone (LHRH) and desmopressin.

## PHYSIOLOGICAL FACTORS

### Nasal Secretions

About 1.5-2.1 ml of mucus is produced daily in the nasal cavity. The mucous exists as a double layer (5 μm thickness). However, the ciliary movement continuously removes this mucous layer. Mucociliary clearance, viscosity of the nasal secretion, solubility of the drugs in the nasal secretions and pH of the nasal secretions are the factors that significantly influence the nasal absorption of the drugs.

### Blood Supply and Neuronal Regulation

Nasal cycles of congestion and relaxation regulate the rise and fall in the amounts of drug permeated, respectively.

## Pathological Conditions

Diseases such as the common cold, rhinitis, atrophic rhinitis, and nasal polyposis are usually associated with mucociliary dysfunctioning, hypo or hyper secretions and irritation of the nasal mucosa, which can influence drug permeation.

## Environmental Conditions

A linear increase in the ciliary beat frequency occurs with an increase in temperature, which in turn influences the properties of the mucous membrane.

## Formulation Factors

The bioavailability of drugs from nasal formulations depends on the physicochemical properties of drug and formulation. Both the aspects work together to yield optimal drug delivery across the membrane.

## Physicochemical Properties of Drug

To be distributed optimally from the nasal formulation, there are certain criteria that the drug should satisfy.

The influence of different physicochemical properties on this distribution is discussed in this section.

## Molecular Weight and Lipophilicity

The absorption of molecules less than 300 Da may not be influenced by their physicochemical properties. However, the diffusion of peptides and proteins greater than 1kilo Da are affected by their molecular weight. The bioavailability of such macromolecules was found to vary from 0.5 to 5%. Lipophilic drugs have been found to be relatively more permeable across the nasal epithelium.

## Solubility

As nasal secretions are more watery in nature, a drug should have appropriate aqueous solubility for increased dissolution. Prodrugs such as testosterone 17 β-N,N-dimethylglycinate hydrochloride have been synthesized to overcome this poor solubility problem.

## Partition Coefficients and pKa

The normal range for pH nasal secretions is about 5.5-6.5. The drug pKa determines the fraction of the drug in un-ionized form under these conditions.

The greater the un-ionized fraction, the higher is the partitioning into the nasal epithelium and the better the absorption.

However, when the absorption of benzoic acid was studied at a pH of 7.19 (99.9% of the drug existed in the ionized form), it was found that greater than 10% of the drug was absorbed, indicating that the ionized species also permeates through the nasal mucosa. However, un-ionized species are absorbed four times faster than the ionized species.

## Physicochemical Properties of the Formulation

The physicochemical parameters of the formulation affect properties of the drug and permeability status of the membrane.

Therefore, it is a challenging task to arrive at the best formulation parameters to achieve therapeutically effective levels of drug delivery.

## pH and Mucosal Irritability

However, the pH of the nasal formulation as well as that of the nasal surface can affect the drug's permeation.

To avoid nasal irritation, the pH of nasal formulation should be adjusted to 4.5-6.5. In addition to avoiding nasal irritation, it also prevents the growth of the bacteria.

## Osmolarity

Drug absorption can be affected by the tonicity of the nasal formulation. Shrinkage of epithelial cells has been observed in the presence of hypertonic solutions. Hypertonic saline solutions also inhibit or terminate cilliary activity, thus reducing mucociliary clearance. However, chronic administration of such solutions may generate pathological implications. Therefore, isotonic solutions are usually preferred for nasal administration.

## Viscosity

High viscosity formulations increase the contact time between drug and nasal mucosa. It also reduces ciliary beating, thus bringing down the mucociliary clearance of the drugs. For example, incorporation of 2% methyl cellulose in a nasal formulation was shown to enhance the relative bioavailability of metoprolol by 40% in rats.

## Dosage Form

Nasal drops are the simplest and most convenient form of dosage but the exact amount delivered cannot be easily quantified and this often leads to over dosage. Moreover, rapid nasal drainage is a problem with drops. Solutions and suspensions sprays are preferred over powder sprays because powder may cause mucosal irritation. Recently, a metered- dosage device was developed that can accurately deliver these nasal drugs. Gels reduce postnasal drip, anterior leakage and localized the formulation in mucosa. Specialized systems such as lipid emulsions, microspheres, liposomes, proliposomes, films, and niosomes have been developed for nasal delivery. These offer better permeation for the drugs as they provide intimate and prolonged contact between the drug and mucosal membrane.

## Device-Related Factors

The size of the droplet or particle delivered and the type of surface distribution does influence the absorption of drugs. The site and pattern of deposition is affected by the formulation composition, physical form of the formulation (liquid, viscous, semisolid, solid etc.), device used, design of actuators and adapters, and administration technique.

***Various Approaches Used for Intranasal Drug Delivery***

Tirucherai et al., has studied a prodrug approach to change the chemical-physical properties of nasally administered drugs with the aim to enhance their ability to cross the nasal mucosa and to prevent their metabolic degradation. Dalpiaz et al. showed that the linkage of biocleavable moieties of drugs can induce a significant increase in their bioavailability. As an example, the synthesis of ester prodrugs allows to easily introduce different functionalities at hydroxyl or carboxylic drug groups, changing their octanol/water partition coefficient and their stability in physiological fluids. Moreover, chemical or enzymatic bioconversion can be controlled by appropriate substitutions. The importance of nasally administered prodrugs was explained by Sasahara et al., less than 1% of orally L-DOPA administered reaches the brain unchanged, being transformed to dopamine during first pass metabolism.

Kao et al., transformed L-DOPA in its butyl ester, giving a prodrug that was more lipophillic and water soluble than the parent compound. The DOPA structure has been indeed changed from a zwitter ionic form to an amine salt. Rat *in vivo* studies substantiated that after nasal administration; this prodrug was rapidly absorbed into the systemic circulation and converted to L-DOPA without significant formation of dopamine. It would appear that the utilization of water soluble prodrugs of L-DOPA via the nasal route may have therapeutic success in the treatment of Parkinson's disease. Similar results were obtained by Wang et al., with ester prodrugs of 17 β-estradiol; a drug that can decrease the risk of Alzheimer's disease, but shows poor water solubility for nasal administration. Unfortunately, the results of oral administration of the 17 β-estradiol proved not to be suitable, since more than 95% of the dose converted to metabolites before reaching the blood. The nasal administration for water soluble 17 β-estradiol resulted in high levels of estradiol in the CSF. Prokai et al., studied how quinols are rapidly converted back to the parent estrogens via an enzyme-catalyzed reduction by using NADPH as a coenzyme and unlike redox cycling of catechol estrogens, without the production of reactive oxygen species. Due to this process, throughout neuroprotection against oxidative stress quinols act essentially as prodrugs for the active hormone.

According to Zhang et al., the microemulsion system is a promising approach for intranasal delivery of Nimodipine for the prevention and treatment of neurodegenerative diseases. Jogani et al., had prepared an intranasal mucoadhesive microemulsion with tacrine to improve brain targeting. Its pharmacokinetic and pharmacodynamic performances for brain targeting were studied and observed for memory improvement in scopolamine-induced amnesic mice. Vyas et al., prepared an intranasal mucoadhesive microemulsion with clonazepam for rapid drug delivery to the brain to treat acute status epileptic patients. Characterization and evaluation were performed for clonazepam microemulsion *in vitro* and *in vivo* in rats.

Intranasal mucoadhesive microemulsion of zolmitriptan was prepared for rapid drug delivery to the brain to treat acute attacks of migraine. Microemulsion was characterized and evaluated for biodistribution in rats by Vyas et al. Vyas et al., demonstrated the rapid and the larger extent of transport for the microemulsion of Sumatriptan compared with the microemulsion of Sumatriptan Succinate; Sumatriptan marketed preparation and Sumatriptan Succinate Solution in the rat brain. Hence, intranasal delivery of Sumatriptan microemulsion developed in this investigation can play a promising role in the treatment of acute attacks of migraine. Surface engineering of nanoparticles with lectins opened a novel pathway to improve the brain uptake of agents loaded by biodegradable PEG-PLA nanoparticles and the following intranasal administration were studied by Gao et al. The UEA I-modified nanoparticles might serve as potential carriers for brain drug delivery, especially for mental therapeutics with multiple biological effects. Poly (ethylene glycol)–poly (lactic acid) (PEG–PLA) nanoparticles was established by Gao et al., in this study. Mucoadhesive nanoemulsions indicated to be more effective and have the best brain targeting for Resperidone. Kumar et al., demonstrated the rapid and the larger extent of transport for Resperidone by Resperidone mucoadhesive nanoemulsions (intranasally) when compared with Resperidone (intranasally) in the rat brain. Tao et al., studied intranasal administration of Huperzine A; this nasal *in situ* gel significantly increased the distributions of Hup A in rat brain tissues, especially for the cerebrum and hippocampus which should be the target areas of Hup A and enhanced the brain targeting of Hup A.

The microparticles have been obtained by spray-drying in the presence of mannitol and lecithin, or chitosan, as carriers. The drug content was detected by UV-spectrophotometer analysis, morphology by scanning electronic microscopy and particle size by laser diffraction. *In vitro* drug release studies from microspheres were achieved according to USP XXIV Nasal administration of microparticles in rats obtained by spray-drying can be performed to obtain the selective CNS targeting of anti-ischemic drugs.

### *Excipients Used to Improve the Absorption of Drugs*

Absorption of polar drugs can be increased by the absorption of promoting agents. Surfactants, bile salts, fatty acids, and phospholipids, have been reported to increase the absorption of nasal administered drugs. These agents work by changing the properties of the phospholipids bilayer structure of epithelial cells that can remove proteins or have the ability to destroy the outer layer of mucosa; so transcellular transport of these drugs will be increased but this can also induce high irritation when inhaled in the nasal cavity.

## Buffers

The lysozome in nasal secretion is responsible for destroying certain bacteria at acidic pH. In alkaline conditions, the lysosome is inactivated and the nasal tissue is hence susceptible to microbial infection. It is therefore advisable to maintain the formulation at a pH of 4.5 – 6.5 bearing in mind the physicochemical properties of the drug, as drugs are absorbed in the un-ionized form.

Due to the small volume of administration, nasal secretions may alter the pH of the formulation. This can affect the concentration of the un-ionized drug available for absorption. Therefore, an adequate formulation buffer capacity is required to maintain the pH *in situ*. Some of the commonly employed buffers include isotonic potassium phosphate buffer, citrate, and acetate buffer.

## Osmotic Agents

The osmolarity of the formulation may be adjusted using sodium chloride. In some cases, sodium sulfite or sodium acid phosphate is used.

## Gelling/Viscosifying Agents or Gel – Forming Carriers

Increasing the viscosity of the solution may prolong the therapeutic effect of nasal preparations. The use of a combination of carriers is often recommended from a safety point of view. Common materials that serve as bioadhesive excipients in the formulation are carbopol, cellulose agents, starch, dextran, chitosan, etc.

## Solubilizers

Conventional solvents or cosolvents such as glycols, small quantities of alcohol, transcutol, medium chain glycerides and labrasol can be used to enhance the solubility of drugs. Other options include the use of surfactants or cyclodextrins such as HP-β-cyclodextrin that serve as biocompatible solubilizers and stabilizers in combination with lipophilic absorption enhancers. In such cases, their impact on nasal irritancy should be considered.

## Preservatives

Depending upon the drug and/or formulation characteristics, sterility may be accomplished with aseptic filling processes, terminal sterilization or both. Sterility must also be maintained throughout a product's shelf life. Unit dose delivery systems or specialized multidose delivery systems may be used to ensure continued product sterility. Alternatively, preservatives may be added to prevent bacterial growth. Parabens, benzylkonium chloride, phenyl ethyl alcohol, and benzoyl alcohol, are some of the commonly used preservatives in nasal formulations. Mercury-containing preservatives have a fast and irreversible effect on ciliary movement and should not be used in nasal systems.

## Antioxidants

Commonly used antioxidants in nasal formulations are sodium metabisulfite, sodium bisulfate, butylated hydroxyltoluene, and tocopherol. Usually, antioxidants do not affect drug absorption or cause nasal irritation. Chemical/physical interaction of antioxidants and preservatives with drugs, excipients, manufacturing equipment and packaging components should be considered during formulation development.

## Absorption Enhancers

Along with the common excipients added to maintain drug solubility, stability, and ideal formulation characteristics, the formulation may require nasal absorption enhancers when the drug is a polar or a macromolecule. The main limiting factor associated with the addition of enhancers to a nasal formulation is the potential toxicity of the nasal mucosa. Nasal absorption enhancers are required to be non-irritating, non-toxic, and non-allergic, or at least to have immediate reversible effects. Moreover, they should be systemically inert in the concentrations used. A large number of absorption enhancers used in combination with drugs increase the permeation of compounds by increasing fluidity of the membrane, decreasing viscosity of the mucosal layer, inhibiting the proteolytic enzymes, distributing the tight junctions, increasing paracellular or transcellular transportation, increasing blood flow, dissociating protein aggregation or initiating pore formulation, or by a combination of these factors. Apart from this, mucoadhesive dosage forms have also been shown to increase the permeation of compounds.

## Surfactants

Surfactants such as bile salts were the first to be tested, and several other promoters have been subsequently investigated. Non-ionic and anionic surfactants including bile salts were found to enhance the nasal absorption of the drugs by multiple mechanisms such as alteration of the mucous layer, opening of the tight junctions between the epithelial cells, reversed micelle formation in the membrane, extraction of membrane components by co-micellization and inhibitory effects on proteolytic enzymes.

## Chitosan

Chitosan is a mucoadhesive material and its mucoadhesive property depends on the positively charged amino group of its D-Glucosamine units. It is capable of increasing the half-life of clearance from 2 to 5 times in the human nasal cavity. Chitosan not only shows bioadhesive properties, but it also opens the epithelial tight junctions that promote the paracelllular permeation of polar drugs across the nasal mucosa. For example, nasally administered insulin in human volunteers can be absorbed at the systemic level only in the presence of chitosan. It has been reported that in particulate systems, in presence of PEG on the surface, can increase the absorption of proteins through the nasal mucosa.

## PLGA

The employment of Poly-(Lactic Acid)-Poly-(Ethylene glycol) (PLA-PEG) in formulation of nano and micro particles are useful to increase the transport of drugs across the nasal mucosa. The high PEG coating density and small size are the optimal conditions that allow it to cross the mucosa.

## PLA

Poly (lactic acid) (PLA) is a biodegradable polymer widely employed to prepare micro and nano particles for brain targeting. As it is biocompatible, it is degraded in body as a lactic acid via crebs cycle, is metabolized without induction of toxic effects.

## Cyclodextrins

Cyclodextrins are used in nasal formulations mainly as a means of providing drug solubilization. However, some reports have demonstrated the permeability enhancement effect of cyclodextrins on nasal mucosa. They are believed to act by increasing drug availability at the formulation – barrier interface.

## Complexing Agents

The integrity of tight junctions, which may be seen as barriers to the paracellular diffusion of molecules, is dependent on extracellular $Ca^{+2}$. It is believed that the mechanism of action of ethylene diamine tetra acetic acid includes depletion of $Ca^{+2}$ from the tight junctional areas, thus allowing the junctions to open. Co-administration of protease inhibitors has been one of the strategies adopted to prevent the metabolism of the drug in the nasal lumen. Often, combinations of permeation enhancers prove synergistic. Abe et al., had studied the pharmacokinetics of LHRH agonist-buserelin on nasal administration in rats. Oleic acid and hydroxylpropyl β-cyclodextrin provided better drug solubility and stability, thus enhancing absorption significantly. The area under the plasma concentration time curve was 33.2±5.5 ng/ml/h when buserelin acetate (0.1 mg/kg) was administered intranasally. The same was 109±11.4 ng/ml/h when oleic acid was administered. The presence of hydroxylpropyl β-cyclodextrin along with oleic acid enhanced the bioavailability of buserelin acetate to 144.3±26.9 ng /ml/h.

## REFERENCES


[1] Illum, L. (2000). Transport of drugs from the nasal cavity to the central nervous system, *Eur. J. Pharm. Sci.*, *11*, 1-18.

[2] Vyas, T. K., Shahiwala, A., Marathe, S. & Misra, A. (2005). Intranasal drug delivery for brain targeting, *Curr. Drug Del.*, *2*, 165-175.

[3] Bodor, N. & Buchwald, P. (1999). Recent advances in the brain targeting of neuro pharmaceuticals by chemical delivery systems. *Adv. Drug Deliv. Rev.*, *36*, 229 (1999).

[4] Talegaonkar, S. & Mishra, P. R. (2004). Intranasal delivery: An approach to bypass the blood brain barrier. *Ind. J. Pharmacol.*, *36*(3), 140-147.

[5] Lu, W., Jiang, W., Chen, J., Yin, M., Wang, Z. & Jiang, X. (2008). Modulation of brain delivery and copulation by intranasal apomorphine hydrochloride, *Int. J. Pharm.*, *349*(1-2), 196-205.

[6] Mathison, S., Nagilla, R. & Kompella, U. B. (1998). Nasal route for direct delivery of solutes to the central nervous system: Fact or fiction? *J. Drug Target*, *5*, 415-441.

[7] Thorne, R. G. & Frey, W. H., II. (2001). Delivery of neurotrophic factors to the central nervous system: Pharmacokinetic considerations, *Clin. Pharmacokinet.*, *40*, 907-946.

[8] Thorne, R. G., Emory, C. R., Ala, T. A. & Frey, W. H., II. (1995). Quantitative analysis of the olfactory pathway for drug delivery to brain, *Brain Res.*, *692*, 278-282.

[9] Balin, B. J., Broadwell, R.D., Salcman, M. & el-Kalliny, M. (1986). Avenues for entry of peripherally administered protein to the central nervous system in mouse, rat, squirrel and monkey, *J. Comp. Neurol.*, *251*, 260-280.

[10] Broadwell, R. D. & Balin, B. J. (1985). Endocytic and exocytic pathways of the neuronal secretory process and trans-synaptic transfer of wheat germ agglutinin-horseradish peroxidise *in vivo, J. Comp. Neurol.*, *242*, 632-650 (1985).

[11] Frey, W. H., Liu, J., Thorne, R. G. & Rahman, Y. E. (1995). Intranasal delivery of 125 I-labeled nerve growth factor to the brain via the olfactory route. In: K. Iqbal, J.A. Mortimer, B. Winblad, H.M. Wisniewski, (eds.), *Research Advances in Alzheimer's Disease and Related Disorders.* John Wiley and Sons Ltd, New York, 329-335.

[12] Frey, W. H., Liu, J., Chen, X., Thorne, R. G., Fawcett, J. R. & Ala, T. A. (1997). Delivery of 125 I-NGF to the brain *via* the olfactory route. *Drug Delivery 4*, 87-92.

[13] Chen, X. Q., Fawcett, J. R., Rahman, Y. E., Ala, T. A. & Frey, W. H. (1998). Delivery of nerve growth factor to the brain *via* the olfactory pathway. *J. Alzheimer's Dis.*, *1*, 35-44.

[14] Frey, W. H., Thorne, R. G. & Pronk, G. (2000). Delivery of insulin like growth factor-1 to the brain and spinal cord along olfactory and trigeminal pathways following intranasal administration: A noninvasive method for bypassing the blood brain barrier, *Soc. Neurosci. Abstract*, *26*, 1365-1370.

[15] Kida, S., Pantazis, A. & Weller, R. O. (1993). CSF drains directly from the subarachnoid space into nasal lymphatics in the rat: Anatomy, histology and immunological significance, *Neuropathol. Appl. Neurobiol.*, *19*, 480-488.

[16] Lowhagen, P., Johansson, B. B. & Nordborg, C. (1994). The nasal route of cerebrospinal fluid drainage in man: A light microscope study. *Neuropathol. Appl. Neurobiol.*, *20*, 543-550.

[17] Foldi, M. (1996). The brain and lymphatic system (I), *Lymphology*, *29*, 1-9.

[18] Hirai, S., Yashiki, T. & Matsuzawa, T. (1981). Absorption of drugs from the nasal mucosa of rat, *Int. J. Pharm.*, *7*, 317-325.
[19] Houna, R. F., Black, A. & Walsh, M. (1971). The deposition of aerosol particles in the nasopharyngeal region of the human respiratory tract, *Aerosol Sci.*, *2*, 47-61.
[20] Bates, D. V., Fish, B. R., Hatch, T. F., Mercer, T. T. & Marrow, P. E. (1966). Deposition and retention models for internal dosimetry of human respiratory tract. Task group on lung dynamics, *Health Phys.*, *12,* 173-207.
[21] Proctor, D. F. Andersen, I. & Lundqvist, G. (1973). Clearance of inhaled particles from humane nose, *Arch. Intern Med.*, *131*, 132-139.
[22] Kublik, H. & Vidgren, M. T. (1998). Nasal delivery systems and their effect on deposition and absorption, *Adv. Drug Deliv. Rev.*, *29*, 157-177.
[23] Mygind, N. & Vesterhauge, S. (1978). Aerosol distribution in the nose, *Rhinology, 16*, 79-88.
[24] Mygind, N. (1973). *Nasal Allergy*, 2nd edit., Oxford: Blackwell.
[25] Khanvilkar, K., Donovan, M. D. & Flanagan, D. R. (2001). Drug transfer through mucus, *Adv. Drug Deliv. Rev.*, *48*, 173-193.
[26] Martin, E., Verhoef, J. C., Cullander, C., Romeijn, S. G., Nagelkerke, J. F. & Merkus, F. W. (1997). Confocal laser scanning microscopic visualization of the transport of dextrans after nasal administration to rats: Effects of absorption enhancers, *Pharm. Res.*, *14*, 631-637.
[27] McMartin, C., Hutchinson, L. E., Hyde, R. & Peters, G. E. (1987). Analysis of structural requirements for the absorption of drugs and macromolecules from the nasal cavity, *J. Pharm. Sci.*, *76*, 535-540.
[28] Agarwal, V. & Mishra, B. (1999). Recent trends in drug delivery systems: Intranasal drug delivery, *Indian J. Exp. Biol.*, *37*, 6-16.
[29] Fisher, A. N., Brown, K., Davis, S. S., Parr, G. D., Smith, D. A. (1987). The effect of molecular size on the nasal absorption of water-soluble compounds in the albino rat, J. *Pharm. Pharmacol.*, *39*, 357-362.
[30] Wheatly, M. A., Dent, J., Wheeldon, E. B. & Smith, P. L. (1988). Nasal drug delivery: An *in vitro* characterization of trans epithelial electrical properties and fluxes in the presence or absence of enhancer. *J. Control Rel.*, *8*, 167-177.
[31] Jones, N. S., Quraishi, S. & Mason, J. D. (1997). The nasal delivery of systemic drugs, *Int. J. Clin. Pract.*, *51*, 308-311.
[32] Corbo, D., Liu, J. & Chien, Y. W. (1989). Drug absorption through mucosal membranes: Effect of mucosal route and penetrant hydrophilicity, *Pharm. Res.*, *6*, 848-852.
[33] Romeo, V. D., De Meireles, J., Sileno, A. P., Pimplaskar, H. K. & Behl, C. R. (1998). Effects of physiochemical properties and other factors on systemic nasal drug delivery, *Adv. Drug Del. Rev.*, *29*, 89-116.
[34] Vila, A., Gill, H., McCalluion, O. & Alonso, M. J. (2004).Transport of PLA-PEG particles across the nasal mucosa: Effect of particle size and PEG coating density. *J. Controlled Rel.*, *98*, 231.
[35] Vila, A., Sanchez, A., Evora, C., Soriano, I., McCallion, O. & Alonso, M. J. (2005). PLA-PEG particles as nasal protein carriers: The influence of the particle size, *Int. J. Pharm.*, *292*, 43.

[36] Soane, R. J., Frier, M., Perkins, A. C., Jones, N. S., Davis, S. S. & Illum, L. (1999). Evaluation of the clearance characteristics of bioadhesive systems in humans, *Int. J. Pharm.*, *178*, 55.
[37] Soane, R. J., Hinchliffe, M., Davis, S. S. & Illum, L. (2001). Clearance characteristics of chitosan based formulations in the sheep nasal cavity, *Int. J. Pharm.*, *217*, 183.
[38] Illum, L. (2003). Nasal drug delivery—possibilities, problems and solutions, *J. Controlled Rel.*, *87*, 187.
[39] Ravikumar, M. N. W., Muzzarelli, R. A. A., Muzzarelli, C., Sashiwa, H. & Domb, A. J. (2004). Chitosan chemistry and pharmaceutical perspectives, *Chem. Rev.*, *104*, 6017.
[40] Gavini, E., Rassu, G., Sanna, V. Cossu, M. & Giuncbedi, M. (2005). Mucoadhesive microspheres for nasal administration of an antiemetic drug, metoclopramide: *In-vitro/ex-vivo* studies, *J. Pharm. Pharmacol.*, *57*, 1.
[41] Fehm, H. L., Perras, B., Smolnik, R., Kern, W. & Born, J. (2000). Manipulating neuropeptidergic pathways in humans: A novel approach to neuro-pharmacology? *Eur. J. Pharmacol.*, *405*, 43.
[42] Illum, L. & Davis, S. S. (1992). Intranasal insulin: Clinical pharmacokinetics, *Clin. Pharmacokinet.*, *23*, 30.
[43] Tobio, M., Gref, R., Sanchez, A., Langer, R. & Alonso, M. J. (1998). Stealth PLA-PEG nanoparticles as protein carriers for nasal administration, *Pharm. Res.*, *15*,270.
[44] Bala, I., Hariharan, S. & Kumar, R. (2004). PLGA nanoparticles in drug delivery: The state of the art, *Crit. Rev. Ther. Drug.*, *21*, 387.
[45] Jalil, R. U. (1990). Biodegradable poly (lactic acid) and poly (lactide-co-glycolide) polymers in sustained drug delivery, *Drug Dev. Ind. Pharm.*, *16*, 2353.
[46] Johansen, P., Men, Y., Merkle, H. P. & Gander, B. (2000). Revisiting PLA/PLGA microspheres: An analysis of their potential in parenteral vaccination, *Eur. J. Pharm. Biopharm.*, *50*, 129-146.
[47] Tirucherai, G. S., Yang, C. & Mitra, A. K. (2001). Prodrugs in nasal drug delivery, *Expert Opin. Biol. Ther.*, *1*, 49-66.
[48] Dalpiaz, A., Scatturin, A., Menegatti, E., Bortolotti, F., Pavan, B., Biondi, C., Durini, E. & Manfredini, S. (2001). Synthesis and study of 5'-ester prodrugs of N6-cyclopentyladenosine, a selective A1receptor agonist, *Pharm. Res.*, *18*, 531-536.
[49] Sasahara, K., Nitanai, T., Habara, T., Morioka, T. & Nakajimai, E. (1980). Dosage form design for improvement of bioavailability of levodopa III: Influence of dose on pharmacokinetic behavior of levodopa on dogs and Parkinson patients, *J. Pharm. Sci.*, *69*, 1374-1378.
[50] Kao, H. D., Traboulsi, A., Itoh, S., Dittert, L. & Hussain, A. (2000). Enhancement of the systemic and CNS specific delivery of L-Dopa by the nasal administration of its water soluble prodrug. *Pharm. Res.*, *17*, 978-984.
[51] Wang, X., He, H., Leng, W. & Tang, X. (2006). Evaluation of brain-targeting for the nasal delivery of estradiol by the microdialysis method, *Int. J. Pharm.*, *317*, 40–44.
[52] Prokai, L., Prokai-Tatrai, K., Perjesi, P., Zharikova, A. D., Perez, E. J., Liu, R. & Simpkins, J. W. (2003). Quinol-based cyclic antioxidant mechanism in estrogen neuroprotection, *PNAS USA 100*, 11741–11746.
[53] Zhang, Q., Jiang, X., Jiang, W., Lu, W., Su, L. & Shi, Z. (2004). Preparation of nimodipine-loaded microemulsion for intranasal delivery and evaluation on the targeting efficiency to the brain, *Int. J. Pharm.*, *275*(1-2), 85-96.

[54] Jogani, V. V., Shah, P. J., Mishra, P., Mishra, A. K. & Misra, A. R. (2008). Intranasal mucoadhesive microemulsion of tacrine to improve brain targeting intranasal mucoadhesive microemulsions of clonazepam: Preliminary studies on brain targeting, *Alzheimer Dis. Assoc. Disorders, 22*(2), 116-124.

[55] Vyas, T. K., Babbar, A. K., Sharma, R. K., Singh, S., Vyas, T. K., Babbar, A. K., Sharma, R. K. & Misra, A. (2005). Intranasal mucoadhesive microemulsions of zolmitriptan: Preliminary studies on brain-targeting, *J. Drug Target, 13*(5), 317-324.

[56] Vyas, T. K., Babbar, A. K., Sharma, R. K., Singh, S. & Misra, A. (2006). Preliminary brain-targeting studies on intranasal mucoadhesive microemulsions of sumatriptan, *AAPS Pharm. Sci. Tech., 7*(1), E1-E9.

[57] Gao, X., Chen, J., Tao, W., Zhu, J., Zhang, Q., Chen, H. & Jiang, X. (2007). UEA I-bearing nanoparticles for brain delivery following intranasal administration, *Int. J. Pharm., 340*(1-2), 207-215.

[58] Gao, X., Tao, W., Lu, W., Zhang, Q., Zhang, Y., Jiang, X., Fu, S. (2006). Lectin-conjugated PEG–PLA nanoparticles: Preparation and brain delivery after intranasal administration, *Biomaterials, 27*(18), 3482.

[59] Kumar, M., Misra, A., Babbar, A. K., Mishra, A. K., Mishra, P. & Pathak, K. (2008). Intranasal nanoemulsion based brain targeting drug delivery system of risperidone, *Int. J. Pharm., 358*(1-2), 285-291.

[60] Tao, T., Zhao, Y., Yue, P., Dong, W. X. & Chen, Q. H. (2006). Preparation of huperzine - a nasal *in situ* gel and evaluation of its brain targeting following intranasal administration, *Yao Xue Xue Bao, 41*(11), 1104.

[61] Zhao, Y., Yue, P., Tao, T. & Chen, Q. H. (2007). Drug brain distribution following intranasal administration of Huperzine A *in situ* gel in rats, *Acta Pharmacologica Sinica, 28*, 273–278.

[62] Ravikumar, M. N. V. (2008). *Handbook of Particulate Drug Delivery*, Vol.2, American Scientific Publishers p. 114.

[63] Weibai, Y. (2008). The nose may help the brain: Intranasal drug delivery for treating neurological diseases, *Future Neurol., 3*(1), 1-4.

[64] Leo, E., Contado, C., Pavan, B., Scatturin, A., Tosi, G., Manfredini, S., Angusti, A. & Dalpiaz, A. (2006). Nanoparticle formulation may affect the stabilization of an anti-ischemic prodrug, *Int. J. Pharm., 307*, 103.

[65] Dalpiaz, A., Leo, E., Vitali, F., Pavan, B., Scatturin, A., Bortolotti, F., Manfredini, S., Durini, E., Forni, F., Brina, B. & Vandelli, M. A. (2005). Development and characterization of biodegradable nanospheres as delivery system of anti ischemic adenosine derivatives, *Biomaterials, 26*, 1299.

[66] Dalpiaz, A., Scatturin, A., Pavan, B., Biondi, C., Vandelli, M. A. & Forni, F. (2001). Poly (lactic acid) microspheres for the sustained release of a selective A1 receptor agonist, *J. Controlled Rel., 73*, 303.

[67] Dalpiaz, A. Scatturin, A., Pavan, B., Biondi, C., Vandelli, M. A. & Forni, F. (2002). Poly (lactic acid) microspheres for the sustained release of anti-ischemic agents, *Int. J. Pharm., 242*, 115-120.

[68] Mathiowitz, E., Lehr, C. M. & Chickering, D. (1998). Bioadhesive formulations for nasal peptide delivery. *Drug Delivery-issues in Fundamentals: Novel Approaches and Development*, Marcel Dekker, New York, 507–539.

*Chapter 3*

# Mucoadhesive Vaginal Delivery Systems

## Vagina as Site for Drug Therapy

The effectiveness of the vagina as a site of drug administration for local effects has been well established. It is an important route for local treatment of several gynecological conditions, such as infections and in hormonal therapy. This route provides advantages such as reducing or eliminating the incidence and severity of side effects, being a non-invasive route of administration and accessibility. These benefits could contribute to a better compliance, thus achieving improved therapeutic outcome. Furthermore, the vagina possesses properties which include: the large surface area of the vaginal wall, permeability, a rich blood supply; and more importantly, the ability to bypass first-pass liver metabolism. These properties are considered to be advantageous in relation to drug absorption.

Currently, there is a variety of pharmaceutical products available on the market designed for intravaginal therapy (tablets, creams, suppositories, pessaries, foams, solutions, ointments, and gels). However, their efficacy is often limited by a poor retention at the site of action due to the self-cleansing action of the vaginal tract. Furthermore, the vagina has unique features in terms of microflora, pH and cyclic changes, and these factors influence the performance of the formulations and must be considered during the development and evaluation of vaginal delivery systems. Therefore, a successful delivery of drugs through the vagina represents a pharmaceutical challenge.

## Anatomy and Physiology of Vagina

The human vagina is a tubular, fibro muscular organ that extends from the cervix of the uterus to the vaginal vestibule, and its measurement in length must be approximated due to individual uniqueness. The vaginal blood supply comes from the internal iliac arteries branching into a complex network of arteries and veins surrounding the vaginal wall. Blood leaving the vagina enters the peripheral circulation via the internal iliac veins, thus bypassing the liver. A vagina histology consists of three distinct layers: an epithelial layer, a middle muscular layer, and an outer fibrous layer. The epithelial layer is classified as a noncornified, stratified squamous epithelium. This layer is composed of lamina propria and an epithelial cell layer which contain particulate glycogen (Figure 1).

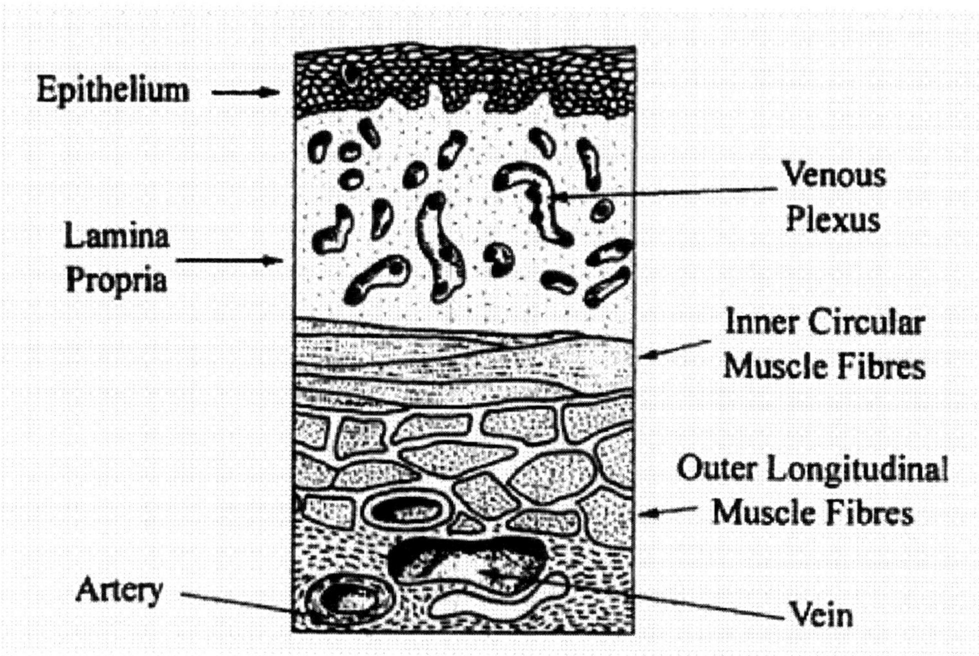


Figure 1. An schematic illustration of the vaginal wall.

The surface of vagina has a number of folds or also called rugae, which increase the surface area of the vaginal wall. Despite of the absence of any glands, the vagina produces a large amount of fluid. This fluid is a mixture of several components and includes leukocytes, inorganic and organic salts, mucins, proteins, carbohydrates, urea, and fatty acids. Like the thickness of the epithelial layer, the amount and composition of the vaginal fluid change upon the hormonal activity. The lactobacilli bacteria are an important component of the vaginal microflora. This bacteria converts glycogen from exfoliated epithelial cells into lactic acid, and as a result, maintains the pH around 4-5, with lowest being around the cervix. Body fluids such as menstrual blood, cervical and uterine secretions will all act as alkalizing agents and increase the vaginal pH.

## Vaginal Environment

The volume, viscosity and pH of vaginal fluid as well as the thickness and porosity of epithelial layer may have either a negative or positive impact on vaginal drug absorption. Therefore, it is crucial to understand the conditions that might influence these parameters. It is noteworthy that the histology and physiology of the vagina may vary with age and with the menstrual cycle. Post-menopausal women experience important changes in the vaginal physiology. These changes manifest as estrogen production declines during the pre-menopause and ongoing menopause. This leads to a permanent decrease in the vaginal glycogen content, and consequently thinning of the vaginal epithelium. Furthermore, elevation of vaginal pH to 6.0-7.5, and a decrease in the quantity of vaginal secretions have been reported. It was estimated that the vaginal secretion produced by postmenopausal women is reduced by 50% compared to that produced by women within the reproductive age. Menstruation is another physiological factor associated with hormonal events. The epithelial layer changes in thickness by approximately 200-300 μm as estrogen levels change throughout the menstrual cycle. Changes in the vaginal pH and viscosity during the menstrual cycle are results of these changes in vaginal histology and physiology. The vaginal pH tends to be lowest at ovulation when estrogen levels reach a peak and both glycogen accumulation and epithelial desquamation at its maximum.

These physiological cyclic variations will be affected by the use of oral contraceptives. Progestin-containing formulations are associated with the production of viscous mucus throughout the treatment cycle. The vaginal pH can be altered by pathological conditions such as infections and inflammations. Diseases such as candidal vaginitis may lower the pH to below 4.5, whereas the inflammatory vaginitis could elevate the pH to over 6.0.

## Changes in Inflammation

The inflammation is a complex reaction representing the host defenses to microbial infection and irritations. Inflammation may take place in two stages, acute and chronic, and mediates tissue repair and regeneration which may occur due to infectious or noninfectious tissue damage. The acute phase of inflammation has a rapid onset and usually lasts for short duration, hours or a few days; and consists of mainly the responses of blood vessels and leukocytes infiltration. Chemical irritations may initiate inflammatory reactions and cause membrane damage of the mucosal epithelial cells, which is followed by the release of prepackaged cytokines. Mediators initiate and amplify the inflammatory response and cause arteriolar dilation, which results in an increase of the blood flow to the injured area. Altering the permeability is another effect of these mediators. Increased permeability results in plasma proteins leaving the vessels through widened interendothelial cell junctions and cause accumulation of protein-rich extravascular fluid. Moreover, circulating leukocytes adhere to the endothelium via adhesion molecules and infiltrate to the site of injury under the influence of chemotactic agents. When acute inflammation is successful in eliminating the offenders, the reaction subsides through activation of endogenous anti-inflammatory response mediated by agents such as interleukin-13 and interleukin-16. However, if the response fails to clear the invaders, inflammation can progress to the chronic phase. Chronic inflammation has longer duration. In this phase leukocytes that are activated by the offending agent and by endogenous mediators may release toxic metabolites and proteases extracellular that can result in damaging tissue. Furthermore, this phase is associated with the presence of lymphocytes and macrophages, the proliferation of blood vessels, fibrosis, causing destruction and remodeling of the tissue. The inflammatory reaction within the female reproductive tract may be essential for immune responses in the clearance of infections and offending agents. However, if this reaction enters the chronic phase and persists, there will be higher probability for complications to arise.

A number of studies have linked chronic inflammation with increased risk of viral infections such as human immunodeficiency virus (HIV). Pro-inflammatory cytokines and the transcription factors controlling the cytokine expression play a major role in HIV-1 pathogenesis. It was reported that continuous stimulation of vaginal epithelial cells by pro-inflammatory mediators could lead to an increase in the availability of potential host cells for infection and viral replication to occur. Furthermore, it also leads to higher probability of viral transmission during sexual intercourse. Several inflammatory agents are also linked to cancer promotion and development. An example of these mediators is the tumor necrosis factor (TNF). TNF induces death of diseased cells at the site of inflammation, and stimulates fibroblast growth. However, if this agent is produced chronically, it may act as an endogenous tumor promoter, contributing to the tissue remodeling and spread tumor growth.

## Vaginal Dosage Forms

The majority of commercially available vaginal delivery systems are usually targeting topical administration. Pessaries (tablets or suppositories) are among the most widely used systems. The principle of their action is that they provide sustained release of the drug as they gradually dissolve or melt. However, this mechanism has a drawback and can result in low bioavailability if the formulation melted faster than intended, thereby giving shorter residence time in the vagina. Vaginal tablets may contain binders, disintegrants and other excipients that are used to prepare conventional oral tablets. Formulating very hydrophobic drugs as vaginal tablets may not be an ideal approach. However, it was suggested that by adding penetration enhancing agents such as surfactants can significantly enhance the drug absorption. Moreover, attempts have been carried out to use mucoadhesive polymers in vaginal tablet formulations in order to increase the residence time. Polyacrylic acid (PAA) is among the bioadhesive polymers that have been utilized for vaginal formulations due to its high bioadhesive strength which allows a longer contact time with the vaginal surface.

Ideally, vaginal drug delivery system that is designed for local effect should distribute uniformly throughout the site of action. However, the distribution and coverage of formulation within the vaginal cavity varies with the properties of the delivery system. It was reported that disintegrating tablet show low coverage whereas solution, suspension, and emulsions display greater distribution profile. Creams and gels are other types of delivery systems frequently used. Creams are normally emulsions whereas gels are usually hydrophilic polymers that utilize covalent bonds to create cross-linked three-dimensional structures. Some formulations such as antifungal emulsion-based formulations seem to have greater advantage over many suppository formulations. An example of gel product is the progesterone gel formulation that is based on a loosely cross-linked poly acrylic acid (Noveon AA1®). This formulation was found to remain on vaginal tissue for 3-4 days, thus allowing dosing intervals of twice a week. However, a disadvantage that can be associated with the use of creams and gels is that they may not provide an exact dose, thus compromising the efficacy of the drug therapy.

## Curcumin

The growing public interest in traditional medicine, particularly plants-based medicine, has led to extensive research on the potentials of natural origin substances. Hundreds of studies were conducted to investigate the effects of natural origin compounds on human health and prevention and treatment of chronic diseases. Among studied compounds, polyphenols appear as one of the most promising groups. In plants, polyphenols are important for growth and protection against pathogens. Polyphenols have recently received much attention in disease prevention and treatment due to their proven antioxidant capabilities. Polyphenols are derived from many components of the human food including peanuts, dark chocolate, green and black tea, and turmeric. Among polyphenols, curcumin is currently one of the most studied substances. It is a hydrophobic, low molecular weight polyphenol widely used in form of the spice, turmeric.

Turmeric has been used in Asia for thousands of years in food, preservation of food, and as a traditional medicine. It is the yellow spice derived from the roots, rhizome, of the plant *Curcuma longa*. The powdered extracts of dried roots, often called turmeric, ukon (in Japanese), or haldi (in Hindi), may contain volatile and nonvolatile oils, proteins, fat, minerals, carbohydrates, moisture, and curcuminoids. The curcuminoids which constitute approximately 5% of most turmeric preparations are a mixture of three principal compounds: curcumin (sometimes referred to as curcumin I), demethoxycurcumin (curcumin II), and bisdemethoxycurcumin (curcumin III). The majority of commercially available curcumin contains the following composition: curcumin I (77%), curcumin II (17%) and curcumin III (3%).

## Delivery Systems for Curcumin

It is necessary to improve the bioavailability of curcumin in order to fully utilize the potential of this agent, and therefore a growing number of research groups are working on this aim. There are studies designed to investigate new approaches that could overcome these limitations seen with free curcumin. A number of studies evaluated the liposomal formulation *in vivo* and their effectiveness. The study conducted investigated the effect of liposomal curcumin on pancreatic carcinoma cells and suppression of KF-kB activity. The incorporated curcumin in liposomes showed a dose related increase in apoptosis of carcinoma cells and suppression of NF- B activity.

Moreover, the liposomal curcumin was found to be as effective as or better than free curcumin. Another experiment had studied the effects of unilamellar liposomal curcumin after tumor implantation in mice. It concluded that liposomal curcumin could increase the life span of the animals by up to 74% in comparison when untreated. The study conducted compared the cellular uptake of liposomal and albumin-loaded-curcumin by the splenic lymphocytes and EL4 lymphoma cells. They reported that liposomes were able to deliver more curcumin into the cells than human serum albumin. The absorption of a micellular formulation was evaluated using everted rat intestinal sacs.

This micellular formulation was composed of phosphatidylcholine and sodium deoxycholate. The authors reported that after the 3 hr incubation, the percentage of free curcumin absorbed was 49%, whereas the percentage for micellular formulation was 56%. In another approach, phospholipid complex of soya phospholipid and curcumin was tested *in vivo* on rats. The study showed higher plasma concentrations, and longer half-life of phospholipid complex in comparison with free curcumin. Furthermore, the bioavailability was also seen to improve significantly after oral administration. The relative bioavailability of curcumin was estimated to be around 30% for the phospholipid complex as compared to free curcumin. Another strategy of delivering curcumin is the self-microemulsifying drug delivery system (SMEDDS). This system is basically composed of isotropic mixtures of oil, surfactant, co-surfactant and a drug which has the ability to form O/W microemulsion when it comes in contact with aqueous medium in the gastro-intestinal tract after oral intake. The curcumin-SMEDDS formulation was composed of 57.5% surfactant, 30% co-surfactant, and 12.5% oil. The *in situ* evaluation of this formulation showed that the absorption percentage of curcumin-loaded SMEDDS was 3.86 times higher than that of curcumin suspension.

"Nanocurcumin" is another formulation recently developed for curcumin. The principle of this formulation is that curcumin is encapsulated in a cross-linked polymeric particle with a hydrophobic core and a hydrophilic shell. The size of these particles lies in the nanometer range and typically less than 100 nm. The group tested the product on pancreatic cancer cells and NFkB and reported to be effective in inhibition of these cells and has similar activity as free curcumin on inflammatory cytokines. Loaded solid lipid nanoparticles (SLN) is another type of nanoparticle based delivery formulations. The system is usually consisting of biodegradable solid lipids. At room temperature the particles are in the solid state. Therefore, the mobility of incorporated molecules is reduced, thus it may offer a possibility for modified release. The study on SLN loaded curcumin was preformed and aimed at using this formulation for topical application. The stability and release was tested and found that properties of cream containing curcumin incorporated into SLNs was improved in comparison to free curcumin in the cream formulation.

Using pharmacological agents such as piperine (a component of black pepper) as suppressor of glucuronidation process of curcumin was also investigated. It was reported the inhibition of this process which occur primarily in the liver and in the intestine could enhance the bioavailability of curcumin. As presented, there are numerous studies suggesting different approaches of delivery systems in order to improve the absorption of curcumin. All of these studies have concluded that it is possible to develop formulations and methods which can improve the bioavailability and give higher plasma concentrations.

A novel mucosal delivery system is described by Ibsen and Uldal. Compositions for the delivery of biologically interactive substances via a mucosal membrane, vaccines, and formulations, as well as for mucosal delivery systems comprising an oxygen-containing metal salt; as in compounds such as: aluminium hydroxide, aluminium phosphate, aluminium sulphate, aluminium acetate, potassium aluminium sulphate, calcium phosphate, calcium tartrate, maalox, beryllium hydroxide, zinc hydroxide, zinc carbonate, barium sulphate or zinc sulphate were described. Furthermore, methods of generating immune responses, vaccinations and treatments of vertebrates, including human beings, as well as methods of treating, preventing or alleviating allergic reactions were also described. A process for preparing the compositions, the vaccines, and the formulations were disclosed. These compositions and delivery systems are useful in a wide range of applications.

Novel vaginal delivery systems as creams, foams, and gels remain for a short time at the application site and repeated administrations are required to obtain effective drug levels. Contrary to this application of mucoadhesive polymers, which extends the drug residence time at the administration site and can solve many of these problems as well as contemporary methods; it also offers a good patient compliance. Mucoadhesive may be defined as the attachment of material/polymer to a mucus coating on the substrate. The vaginal route appears to be highly appropriate for mucoadhesive drug delivery systems. It includes systems for treating largely local conditions, or for use in contraception. The main advantage of mucoadhesive systems is the possibility of increasing the time of residence *in situ*, thus reducing the number of applications for the intended end results. The formulation is expected to be retained at the biological surface and the drug to be released close to the absorptive membrane, with a rise in probability of enhanced bioavailability. Mucoadhesive polymers have been utilized in many different dosage forms in an effort to achieve local, as well as systemic delivery of drugs through the different mucosa. To serve as mucoadhesive, the polymer should possess certain general physicochemical features such as predominantly anionic hydrophilicity with numerous hydrogen bond forming groups, suitable surface property for wetting through mucus/mucosal tissue surfaces and sufficient flexibility to merge with and penetrate the mucus network or tissue crevices. Diverse classes of polymers have been investigated for potential use as mucoadhesives in the vagina. These include synthetic polymers such as poly (acrylic acid), cellulose derivatives, PVA as well as naturally occurring polymers such as hyaluronic acid, chitosan, gelatin, and starch. Partial deacetylation of chitin results in the production of chitosan, which is a polysaccharide comprising copolymers of glucosamine and N-acetyl glucosamine. Chitosan is cationic polyamine with a high charge density at $pH<6.5$, and adheres to negatively charged surfaces. It is also known to exhibit antimicrobial activity. Chitosan is commonly used to obtain pharmaceutical systems for mucosal delivery. Chitosan is currently receiving a great deal of attention for medical and pharma applications, particularly because of its penetration enhancement capability, alongside with other intrinsic properties such as biocompatibility, biodegradability, bioadhesivity, and bacteriostatic effects. Chitosan citrate, in one study, has been evaluated as a multifunctional polymer for vaginal delivery. It was found that, chitosan citrate possess potential properties of penetration enhancement and protease inhibition. In a vaginal delivery system of metronidazole, the possible use of chitosan in mixtures of different ratios with anionic polymers for the preparation of mucoadhesive tablets has been evaluated. By introducing the thiol group on the chitosan, it was found that bioadhesiveness of the chitosan could be significantly increased. In Clotrimazole therapy, controlled release of the Clotrimazole was found to be dependent on the amount of the covalently attached thiol group.

Poly vinyl alcohol is water soluble synthetic polymer represented by the formula $(C_2H_4O)_n$. The value of n for commercially available materials lies between 500 and 5,000, equivalent to a molecular weight range of approximately 20,000 to 200,000. In the development of the bioadhesive vaginal films of PSS, effect of molecular weight of PVA on physical and mechanical properties of films was studied using three different graded, via low MW (300,000-700,000), medium MW (89,000-98,000), and high MW (125,000). Films obtained with all the three grades possessed good peel ability, flexibility, to include both physical and mechanical properties. Bioadhesive strength of PSS films were compared with commercial products VCF and Orto-option and it was found significantly greater for PSS films.

Gelatine is the generic term for a mixture of purified protein fraction obtained either by partial acid hydrolysis (Type A) or by partial alkaline hydrolysis (Type B) of animal collagen. Gelatine may also be a mixture for both types. The old vaginal dosage forms which were based on gelatin can be seen as the first mucoadhesive systems. Medicated tampon for intravaginal sustained administration of metronidazole contains gelatin as mucoadhesive polymer. Gelatin capsules with different mucoadhesive polymers can be suggested.

Polyacrylate and poly (acrylic acid): In most of the vaginal preparations poly (acrylic acid) (PAA) derivatives have been used as mucoadhesive polymer as it has been considered as a good mucoadhesive, among all PAAs. In a study of post-menopausal women, a bioadhesive formulation, Replens gel (one of three percent polycarbophil gel) was shown to retain on vaginal tissue for three to four days. However, conflicting results were obtained when the same formulation was studied in human vagina. There was a lack of significant retention of the gel in five out of six volunteers studied. Polycarbophil gel is able to alter vaginal and suburethral blood flow favorably from the surface of the vaginal tissue, which led to a significant improvement in dry vagina and menopause related stress incontinence. For systemic delivery, insulin suspended in a PAA gel base was observed to facilitate the rate of vaginal absorption in alloxan diabetic tablets in a modified starch-polyacrylic acid mixture which showed an increased potential for curing bacterial vaginosis in women.

Carbopol has a very high molecular weight polymer for an acrylic acid, and has been used mainly for its mucoadhesive properties in vaginal drug delivery systems. Carbopol is used in the formulation of mucoadhesive polymeric films and was developed as female controlled drug delivery system (FcDDS). It was found that, carbopol is able to not only increase the degree of hydration and mucoadhesiveness, but also maintain the morphology of the films in the vaginal cavity. The ability of carbopol gel to attach to lymphocytes also makes them a site-specific drug delivery system for AIDS prophylaxis. FcDDS in a form of carbopol gel has been developed as an intravaginal formulation, whose adhesiveness is controlled by the carbopol concentration. It was reported that higher viscous bioadhesive polymer gel enhanced the duration of the mucosa contact, while the release rate of a drug molecule from polymer based gel remained as fast, as it remained in water. Also, carbopol gel formulations for acyclovir acts as an antiviral agent with low bioavailability especially in topical dosage forms. Acyclovir was encapsulated in three types of liposomes, namely neutral, negatively- and positively- charged liposomes. However, even with incorporation of different liposomes in hydrogels the results showed that the slower released profile of the drug from a liposomal gel in comparison with liposomal suspension, as well as its protective effect of a hydrogel matrix on liposomes yielded in improvement of stability.

Sodium alginate consists of sodium salt in an alginic acid, which is a mixture of polyuronic acid composed of residue of D-mannuronic and L-guluronic acid. The adhesiveness of the hydrogel of sod alginate has been investigated and it is found that adhesiveness increased with concentration. In vaginal delivery of metronidazole tablets, it was found that, the values of adhesion force were more or less constant in the tablets containing 24 % of sod alginate, showing a very slight decrease with the increase in applied weight. Alginate was then cross-linked with calcium chloride to contain3% of the alginate-N-9 and manufactured over a pH range of 3.4-5.9 and at different osmolarity. It was found that delivery gel pH had a significant effect on spermicidal efficacy of the alginate-N-9 system, biodiffusion increased with a decreasing pH.

The USP NF 23 describes carrageenan as the hydrocolloide obtained from some members of class Rhodophyceae (red seaweed) by extraction with water and aq alkali. It consists of potassium, sodium, calcium, magnesium, and ammonium sulphate esters of galactose and 3,6-anhydrogalactose copolymer. Carageenan might be divided into three families according to the position of sulphate group and the presence or absence of anhydrogalactose *viz* lambda, iota, and kappa.

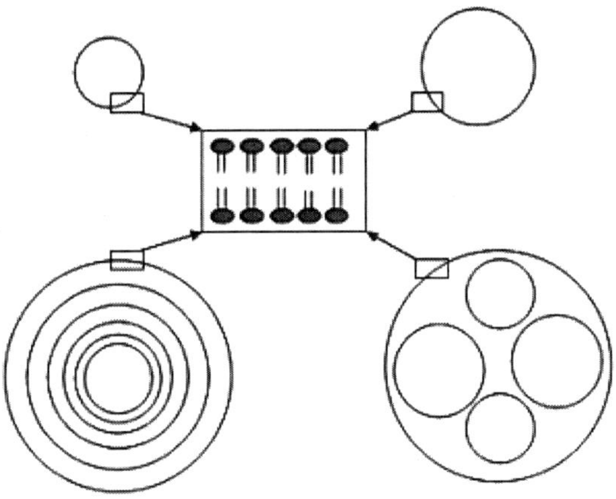


Figure 2. An schematic illustration of liposomes for different sizes.

Various cellulose derivatives like hydroxyl ethylcellulose (HEC), hydroxyl propyl cellulose (HPC), hydroxyl propyl methyl cellulose (HPMC), methyl cellulose (MC), sodium carboxymethyl cellulose (sod CMC) have been studied for intravaginal drug delivery systems. Gynol II, a contraceptive jelly, contains sodium carboxymethyl cellulose as a mucoadhesive polymer and is used as a spermicidal contraceptive in conjunction with barrier methods of contraception. HPMC and HEC were also evaluated as mucoadhesive polymers in the formation of the bioadhesive vaginal films of sodium polystyrene sulfonate (PSS), a novel contraceptive antimicrobial agent. In the study of bioadhesive acyclovir vaginal tablets, the units were prepared by using PAA, MC, CMC, HPC, and HPMC as bioadhesive polymers in different concentrations by direct compression as well as wet granulation techniques. The swelling of tablets containing HPC, CMC, and MC was very rapid and caused the tablets to distintegrate. The tablets with HPMC had swelling behavior that lasted six hours in the lactic acid solution. The maximal detachment force from the vaginal tissue was found to depend on the concentration and type of the bioadhesive polymer. The tablets containing HPMC needed the highest detachment force.

## CLASSIFICATION OF LIPOSOMES

Liposomes are usually classified according to their lamellarity and size. The following categories show the major types of liposomes: Multilamellar vesicles (MLV): This population has a broad range of size distribution that occurs in a range of 100-1000 nm. The lipid composition may influence the lamellarity of these MLVs. However, the lamellarity typically varies between 5 and 20 concentric lamellae. Large unilamellar vesicles (LUV): The size of these vesicles is normally up to 1000 nm and the structure consists of a single lamellae. Small unilamellar vesicles (SUV): The structure normally consists of single lamellae and the diameter of this population is below 100 nm. The choice of preparation method can influence the size and lamellarity of liposomes, resulting in different characteristics and the fate of the entrapped *in vivo* drug.

There are various methods to produce different types of liposomes. However, all preparation methods can be simplified as to involve three basic steps: 1) Preparation of the dispersion of the lipids in aqueous media; 2) purification of prepared liposomes; and 3) analysis of the final product. The following is a brief description in the methods that are considered among the most widely used in liposome preparation: Hand-shaking, (MLV) The principle of this method is to dissolve the lipid in efficient solvent. Lipophilic drugs can be dissolved in a compatible solvent which can be added later to the lipid solution. This mixture will be dried under pressure until it forms a lipid film. Hydrophilic drugs can be dissolved in an aqueous solution which will be added to the lipid film under shaking. This will produce swelling and peeling of the lipid films and gives a milky suspension of MLVs. Using this technique, one can entrap as high as 100% of the lipid soluble molecules, whereas hydrophilic compounds are often encapsulated in amounts of 5-10%. Sonication, (SUV) Sonication is the most widely used method for producing small vesicles. It is usually used to convert MLVs to SUVs through the employment of energy at high levels by the exposure of MLVs to ultrasonic irradiation. Probe-sonication and bath sonication are frequently used techniques to reduce the size of liposomes. The probe has the most efficient transfer of energy to the liposomal dispersion. However, it is associated with metal particle shedding from its probe and therefore one must be aware of the potential contamination. Bath sonication is much milder than probe sonication, but it is time-consuming and may result in low yield of smaller liposomes. Extrusion, (LUV) The principle in extrusion technique is based on employment of moderate pressure to force MLVs through polycarbonate filters with defined pore size. At low pressures (100 psi), MLVs display a reduced-size while maintaining their multi-lamillarity, whereas at higher pressure the liposomes are broken down as they pass though the membrane filter resulting in reorganizing of the phospholipid bilayer giving rise to unilamellar vesicles. It is a simple method and easy to use and there are several products available on the market. Among the equipments are the Hamilton® syringes. These are two syringes connected by a filter holder allowing the samples to pass back-and-forth through the polycarbonate filter. However, this method can be limited by the production of small volumes of LUVs, and the back pressure that can be tolerated when the syringe is used and filter holder is limited.

Unprocessed MLVs have limited uses with *in vivo* studies because of their large diameter and heterogeneity of size. However, the techniques used to change these parameters may influence physical properties of liposomes. The conversion to SUVs from MLVs may result in vesicles with very low trapped volumes. Furthermore, SUVs can be unstable and prone to fusion process due to the high curvature of the lipid bilayer. Extrusion technique used to produce LUVs may result in rupturing and resealing which leads to further leakage of the entrapped drug and the final vesicles may have a lower amount of entrapped material, depending on the lipophilicity of the drug.

## Liposomes as Drug Delivery System

A liposome is defined as a self-forming structure consisting of one or more concentric spheres of lipid bilayers separated by water or aqueous buffer compartments. Phospholipids are the backbone of these structures. Phosphatidylcholine (PC), also called lecithin, is a biocompatible phospholipid that exists in plants and animals and used frequently in liposomal preparation. Moreover, there are other molecules widely used in combination with phospholipids, such as cholesterol. The exact location of a drug in liposomes will depend upon its physicochemical characteristics and the composition of the lipids. However, as a general rule, the hydrophilic drug molecules can be encapsulated in the aqueous space whereas the hydrophobic and amphiphilic molecules can be incorporated into the lipid bilayer.

Numerous studies have provided evidence to demonstrate the ability of liposomes to enhance the efficiency of drug delivery via several routes of administration. One of the major effects of liposomes as drug carriers is altering the pharmacokinetics of a drug. It is known that pharmacological response is dependent upon the concentration of the drug in the target cell. The drug concentration in the target site is governed by absorption, distribution, and elimination. These processes may influence the pharmacokinetics of the drug and lead to inefficient utilization of the therapeutic agent. Thus, higher doses need to be administrated. Furthermore, higher drug doses often lead to resistance and undesirable immunological and toxicological effects. Liposomes are thought to shield all or most of the drug molecules resulting in the decrease in the direct drug contact with the biological environment. Thus, the pharmacokinetic profile of the drug will be determined by the physiochemical properties of liposomes, rather than the drug itself. Incorporating the drug into a vehicle capable of delivering it intact would overcome many of the disadvantages of the free drug administration. Improving the pharmacokinetics of the drug by this method could lead to beneficial effects such as reduced dosages, increased cellular permeability, and delayed drug elimination. It is worth mentioning that liposomes are also non-toxic, biodegradable, and can be manufactured on large scales. The potentials of liposomes to serve as delivery systems have been proven by the number of liposomal formulations already approved by FDA for clinical use such as AmBisome® (amphothericin B) and DaunoXome® (Daunorubicin). These formulations were clinically compared with conventional drug formulations and proved superior for liposomal delivery.

## Topical Administration

Liposomes have been widely used to enhance the efficiency of drug delivery though various routes of administration and have been shown to be significantly superior to conventional dosage forms; especially for intravenous and topical administration. However, therapeutic applications of systemically administered liposomes have been limited by their rapid clearance from the bloodstream and their uptake by reticuloendothelial system (RES) in liver and spleen. Furthermore, the use of liposomal formulations in oral administration has been limited due to physiological factors. The three major factors are pH, bile salt, and pancreatic enzymes in the gastrointestinal (GI) tract can destabilize the structure of the vesicles and limit their potential. Topical liposomal administration might offer an opportunity for developing a novel delivery system that could overcome these limitations experienced with the systemic and oral liposomal formulation as well as conventional products. The major advantages of topical liposomal drug include: 1) Reduction of side effects and incompatibilities that may arise from undesirably high systemic absorption of drug; 2) Markedly increasing the liposomal drug accumulation in the desired tissues; and 3) Capability for incorporation of a wide variety of hydrophilic and hydrophobic drugs. Additionally, their ability to provide a sustained/controlled release and an enhancement of the cellular penetration of the incorporated material could improve their potential for being applied vaginally.

## VAGINAL APPLICATION

Liposomes ability to encapsulate both the lipophilic and hydrophilic active ingredients represents a promising delivery system in regard to vaginal therapy. Some studies have investigated the release profiles for a number of pharmaceutical agents to incorporate liposomal formulations in simulated vaginal conditions. The study conducted aimed to investigate the stability of liposomal formulations for metronidazole, clotrimazole and chloramphenicol, drugs which are frequently applied in the treatment of vaginal infections. The group concluded that the liposomes retained approximately 28-40% of the entrapped drug even after 6 hours of incubation in an environment that mimics the vaginal cavity of pre- and post-menopausal women.

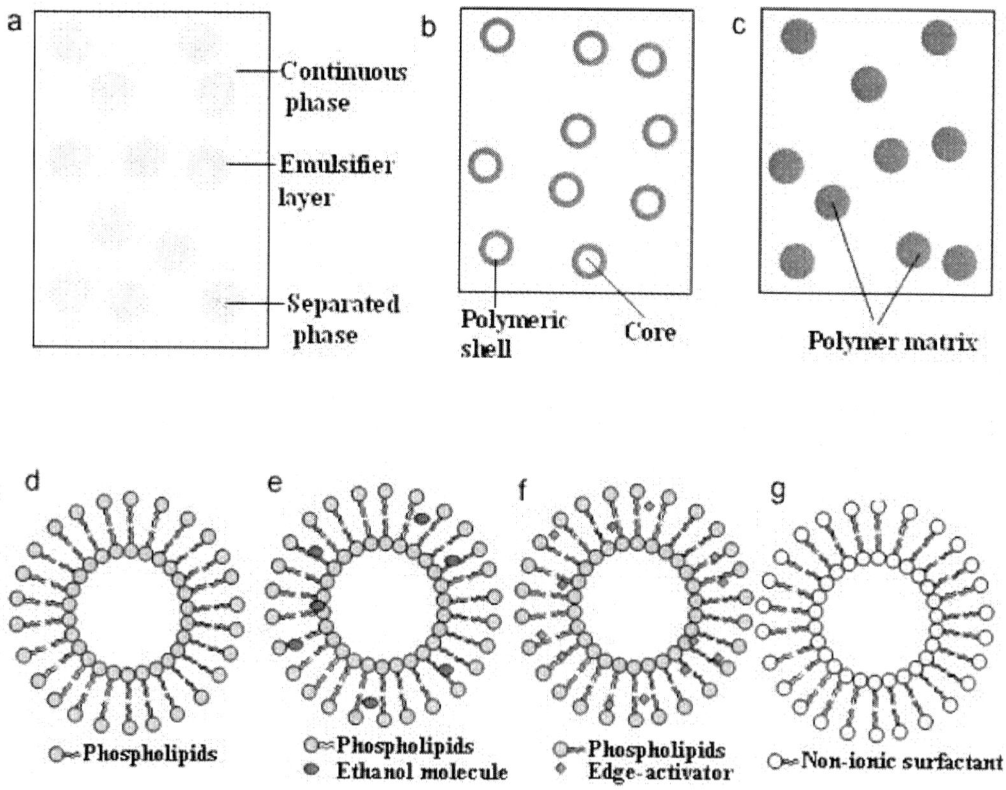


Figure 3. Schematic representation of (a) a nanoemulsion, (b) polymeric nanocapsules, (c) polymeric nanospheres, (b) individual liposomes, (e) an ethosome, (f) a transfersome, and (g) a noisome.

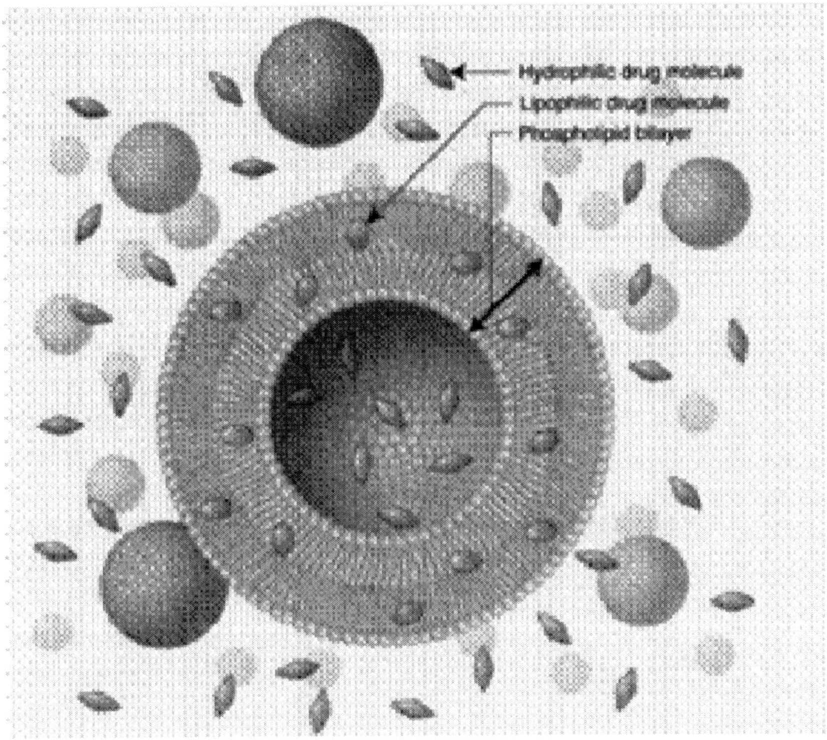


Figure 4. Schematic illustration of small unilamellar liposome-drug carrier with hydrophilic drug in the aqueous compartments and lipophilic drug incorporated into the phospholipid bilayer.

The cervical mucus present in the vagina is believed to assist in the bioadhesion process. However, being in a liquid form as liposomal suspensions, there might be higher probability for the formulation to be expelled from the vagina and reduce the retention time at the site of action. Therefore research groups are considering the incorporation of liposomes as a bioadhesive base, such as hydrogels. Among various types of hydrogels, Carbopol hydrogels have demonstrated a good compatibility with liposomal formulations. The polymer adhesion to tissues permits intimacy of contact and also improves the drug absorption.

Furthermore, it also prolongs the residence time at the site of administration. One study investigated the comparison of the release profiles between liposomal gels containing hydrophilic marker substance (FITC-dextrans) with the same gel with the marker present in non-liposomal form. The conditions in this study were set to be close to the physiological environment. The retention of encapsulated hydrophilic model compound FITC-dextrans was much higher than the control gel without incorporated liposomes. It was found that about 20% of dye was released from liposomes incorporated in the gel after 72 hours, whereas the control gel released approximately 89% during the same period. Another study was investigating the effect of Carbopol gel on the stability of liposomal chloramphenicol. It was calculated that more than 40% of the entrapped drug was in the liposomal gel even after 24 hours of incubation in simulated vaginal conditions. The authors concluded that the gel formulation provided a stable vehicle suitable for vaginal application in which liposomes are distributed uniformly and their original size distribution of liposomes is preserved.

TCS are gaining popularity because of their high mucoadhesiveness and extended drug release properties. The derivatization of the primary amino groups of CS with coupling reagents bearing thiol functions leads to the formation of TCS. The use of MET, which inhibits estrogen biosynthesis, is an attractive treatment for postmenopausal women with hormone dependent cancer. Since the early 1980s, the concept of mucoadhesion has gained considerable interest in pharmaceutical technology. This might open the door for novel, highly efficient dosage forms especially for oral drug delivery. The most important goal in cancer chemotherapy is to minimize the exposure of normal tissues to drugs while maintaining their therapeutic concentration in tumors. Interestingly, nanoparticles (NPs) exhibit a significant tendency to accumulate in a number of tumors after intravenous injection. Hence, uptake and consequently bioavailability of the drug may be increased and frequency of dosing reduced with the result that patient compliance is improved. Various natural and synthetic polymers have been discovered as mucoadhesive excipients. Their mucoadhesive properties can be explained by their interaction with the glycoproteins of the mucus, based mainly on non-covalent bonds such as ionic interactions, hydrogen bonds, and van der Waals forces. The biopolymer CS is obtained by alkaline deacetylation of chitin which is one of the most abundant polysaccharides in nature. Shell wastes of shrimp, lobster and crab are the main industrial sources of chitin. The primary amino group accounts for the possibility of relatively easy chemical modification of CS and salt formation with acids. At acidic pH, the amino groups are protonated, which promotes solubility, whereas CS is insoluble at alkaline and neutral pH. Because of its favorable properties, such as enzymatic biodegradability, non-toxicity and biocompatibility CS has received considerable attention as a novel excipient in drug delivery systems, and has been included in the European Pharmacopoeia since 2002. The administration of nanoparticles will also provide the advantage of facilitating their injection through standard infiltration needles. So far, there was one paper published on MET nanoparticles that had a direct precipitation technique. Recently, it has been shown that polymers with thiol groups provide much higher adhesive properties than polymers generally considered to be mucoadhesive. To increase patient compliance, to overcome the undesirable side effects, MET could be entrapped into biodegradable nanoparticles for sustained delivery; so that, it can inhibit estrogen biosynthesis for a prolonged time by virtue of increased local concentration of the drug at the receptor site. To date, three different TCS derivatives have been synthesized: CS-thioglycolic acid conjugates, CS-cysteine conjugates, and chitosan-4-thio-butyl-amidine (CS-TBA) conjugates. These TCS have numerous advantageous features in comparison to unmodified CS, such as significantly improved mucoadhesive and permeation enhancing properties. The strong cohesive properties of TCS make them highly suitable excipients for controlled drug release dosage forms. We have prepared MET-loaded TCS–PMAA nanoparticles by emulsion solvent evaporation technique to obtain smaller particle size with high entrapment efficiency and sustained release profile. Particle size, morphology, entrapment efficiency, drug–polymer interaction and *in vitro* release of MET–TCS–PMAA nanoparticles were evaluated. The drug percentage influence (related to polymer mass) on formulation performance including particle size, entrapment efficiency and, *in vitro* release were investigated. These nanoparticles had good spherical geometry. It is obvious that the surface of the nanoparticles shrank and a densely cross-linked gel structure was formed. This may explain the greater retardation of drug release in matrices with a higher crosslink content (Scheme 1).

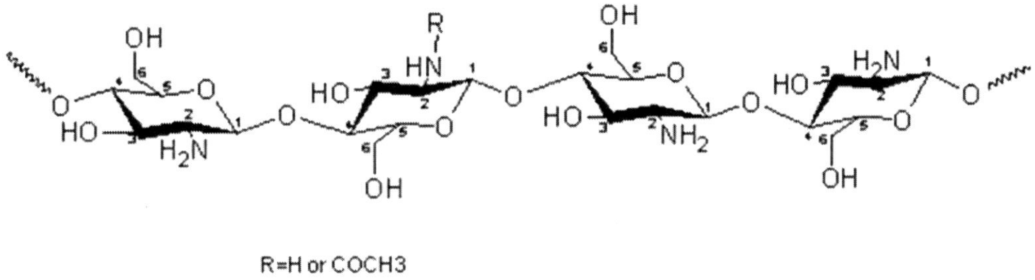


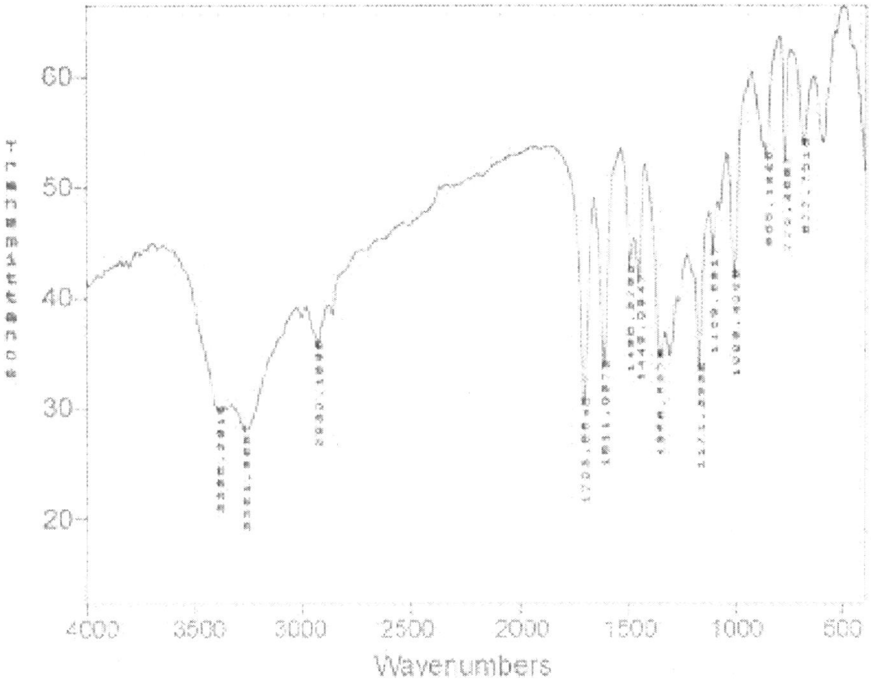


Figure 5. FT-IR spectra of TCS-PMA nanoparticles.

Results of drug content and EE demonstrated that drug content increased from 12.03±0.10 mg/100mg to 28.14±0.03 mg/100mg for nanoparticles when the amount of drug increased from 5% to 20% (wt/wt). No signification increase in drug content was observed further when the amount of drug had increased, *i.e.*, above 15% (wt/wt), which could be due to the limited solubility of the drug in DMF and that is endorsed from the presence of drug particles on the surface of the nanoparticles prepared with 20% of drug concentration. The percent EE was increased up to 87.20±4.00% with increasing polymer concentrations to 4%. Concentration of the cross-linking agent exhibited no significant effect on percent EE. The synthesized polymer was characterized by IR spectroscopy. The FT-IR spectrum of TCS–PMAA is shown in Figure 5. The spectra of TCS–OH and–NH stretch were clearly seen at 3385.292 and 3261.866cm$^{-1}$, respectively. Additional presence of amidine I and amidine II bonds are seen at 1703.855 and 1611.087cm$^{-1}$ corresponding to $\geq NH_2{}^+$ stretch and $NH \geq NH_2{}^+$, respectively, being two coupled vibrations. The presence of an additional bond at 1495.3783 and 1449.0847cm$^{-1}$ can be assigned for N–H bond of the salt $NH_2{}^+Cl-$. Other characteristic peaks of CS O–H stretch, C–H stretch, and C–O stretch were present at 3400–3600, 2930, and 1009–1171cm$^{-1}$, respectively. This confirmed the synthesis of TCS. The spectrum of TCS was well correlated with reports by Matsuda et al., for TCS. Thiol content of TCS was found 214±52_mol/g.

Figure 6 shows the percentage of swelling for TCS–PMAA nanoparticles at different time intervals. The results revealed that nanoparticles swelled rapidly when immersed in phosphate buffer (pH 6.8). It was reported that adhesive properties and cohesiveness of mucoadhesive polymers were generally affected by their swelling behavior. Mucoadhesive nanoparticles are anticipated to take up water from the underlying mucosal tissue by absorbing, swelling and capillary effects, leading to considerable stronger adhesion.

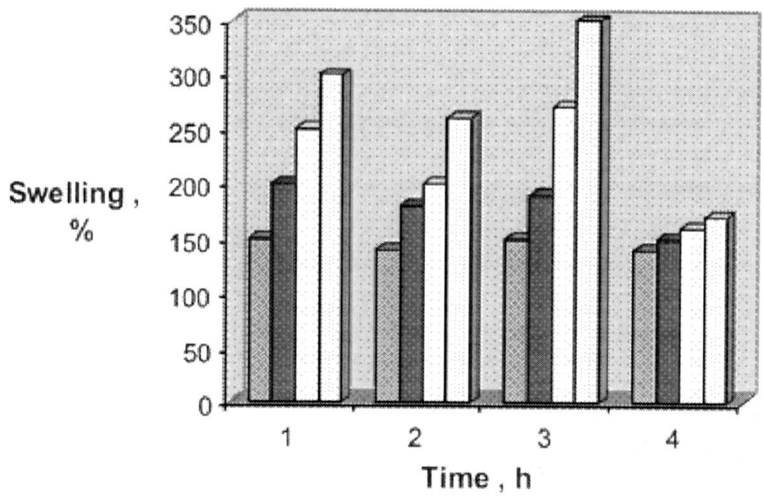


Figure 6. Percentage results for the swelling measurement of TCS-PMA nanoparticles.

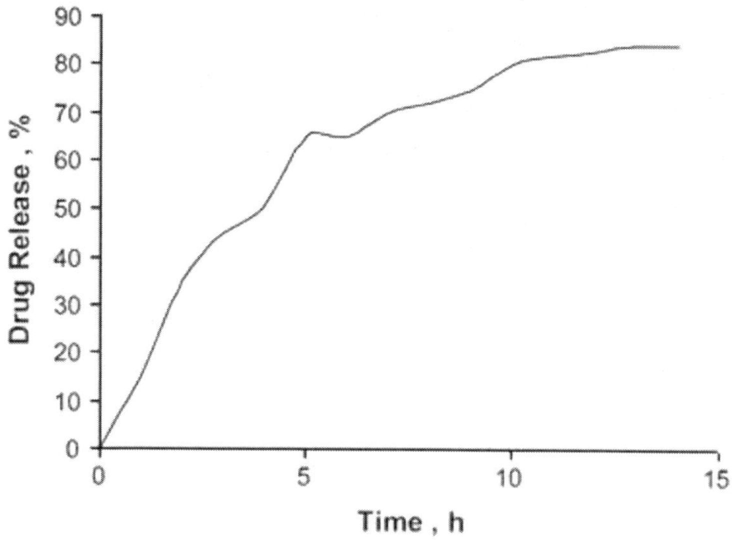


Figure 7. Percentage in vitro drug release of MET from TCS-PMA nanoparticles.

The percent of swelling for the nanoparticle was found to follow the rank order 248±18%, 198±15%, 279±26%, and 164±15%, respectively. It was observed that TCS–PMAA nanoparticles swelled slowly and produced higher mucoadhesive strength. This is perhaps because slow swelling avoids the formation over a hydrated structure that loses its mucoadhesive properties before reaching the target.

Figure 7 shows the release of MET from mucoadhesive nanoparticles. Drug powder was completely released (95.3±4.1%) within 1 hr. The time taken to release 75% of MET from TCS–PMAA nanoparticles was 5.0±0.6, 9.6±1.0, 4.2±0.2, and 5.3±0.5 hr, respectively. The significantly higher time required by the TCS–PMAA nanoparticles to release MET was due to its better stability in acidic medium, which contributed significantly to a smaller amount of drug release during the initial 1 hr of dissolution (30.1±1.5% and 21.2±0.6%) drug released from the TCS–PMAA nanoparticles. This may be attributed to the higher solubility of CS in acidic medium. CS is soluble in acidic medium but cross-linking with glutaraldehyde through its amino group stabilized the nanoparticles matrix and provides the sustained release. The significantly smaller drug release from TCS–PMAA nanoparticles is due to the presence of disulphide bonds in nanoparticle matrix further stabilized the structure along with glutaraldehyde as the cross-linking agent.

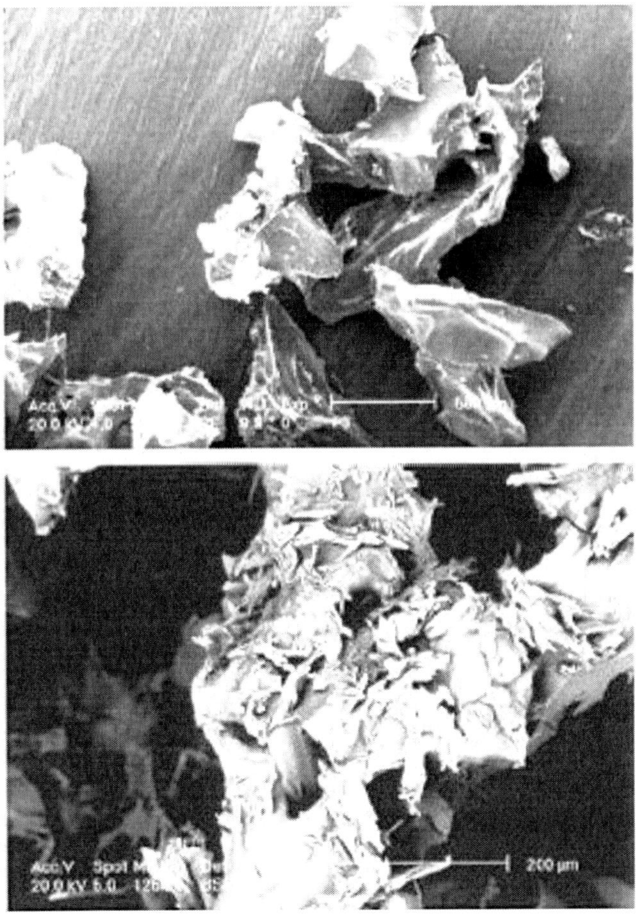


Figure 8. An SEM photomicrograph of TCS-PMA nanoparticles with MET drug.

The morphology of nanoparticles was examined by scanning electron microscopy (SEM, Philips XL-30 E). The nanoparticles were mounted on metal stubs using double-sided tape and coated with a 150 °A layer of gold under vacuum. Stubs were visualized under a scanning electron microscope. Figure 8 shows the SEM of TCS–PMAA nanoparticles that were synthesized by chemical reaction. This nanoparticle is very sensitive to the temperature due to the interaction electron and samples. Scanning electron micrography images were obtained from a diluted solution of the MET particle. The white spots are MET nanoparticles. The SEM image shows the presence of MET spherical particles in polymer matrix, which are homogeneously distributed throughout the polymer, which is also confirmed from 1H NMR studies. As observed from SEM photomicrographs, the crystals of MET have a different appearance than recrystallized MET. These nanoparticles do not have clearly defined crystal morphological features in the SEM photomicrographs.

## REFERENCES

[1] Liu, W. C. & Yao, K. D. (2002). Chitosan and its derivatives a promising non-viral vector for gene transfection, *J. Controlled Rel.*, *83*, 1–11.

[2] Synder, G. H., Ready, M. K., Cennerazzo, M. J. & Field, D. (1983). Use of local electrostatic environments of cysteines to enhance formation of a desired species in a reversible disulfide exchange reaction, *J. Biochem. Biophys. Acta*, *749*, 219–226.

[3] Thanou, M., Nihot, M. T., Jansen, M., Verhoef, J. C. & Junginger, J. C. (2001). Mono-N-carboxymethyl chitosan (MCC), a polyampholytic chitosan derivative, enhances the intestinal absorption of low molecular weight heparin across intestinal epithelia *in vitro* and *in vivo*, *J. Pharm. Sci.*, *90*, 38–46.

[4] Baumann, H. & Faust, V. (2001). Concepts for improved regioselective placement of O-sulfo, N-acetyl and N-carboxymethyl groups in chitosan derivatives, *J. Carbohydr. Res.*, *331*, 43–57.

[5] Andreas, B. S. & Krajicek, M. E. (1998). Mucoadhesive polymers as platforms for peroral peptide delivery and absorption: Synthesis and evaluation of different chitosan–EDTA conjugates, *J. Controlled Rel.*, *50*, 215–223.

[6] Andreas, B. S. & Hopf, T. E. (2001). Synthesis and *in vitro* evaluation of chitosan–thioglycolic acid conjugates, *J. Sci. Pharm.*, *69*, 109–118.

[7] Hornof, M. D., Kast, C. E. & Andreas, B. S. (2003). *In vitro* evaluation of the viscoelastic behavior of chitosan–thioglycolic acid conjugates, *Eur. J. Pharm. Biopharm.*, *55*, 185–190.

[8] Andreas, B. S., Hornof, M. & Zoidl, T. (2003). Thiolated polymers–thiomers: Modification of chitosan with 2-iminothiolane, *Int. J. Pharm.*, *260*, 229–237.

[9] Roldo, M., Hornof, M., Caliceti, P. & Andreas, B. S. (2004). Mucoadhesive thiolated chitosans as platforms for oral controlled drug delivery: Synthesis and *in vitro* evaluation, *Eur. J. Pharm. Biopharm.*, *57* (1), 115–121.

[10] Langoth, N., Guggi, D., Pinter, Y. & Andreas, B. S. (2004). Thiolated chitosan: *In vitro* evaluation of its permeation properties, *J. Controlled Rel.*, *94* (1), 177–186.

[11] Kast, C. E., Valenta, C., Leopold, M. & Andreas, B. S. (2002). Design and *in vitro* evaluation of a novel bioadhesive vaginal drug delivery system for clotrimazole, *J. Controlled Rel.*, *81*, 347–354.

[12] Leitner, V. M., Marschutz, M. K. & Andreas, B. S. (2003). Mucoadhesive and cohesive properties of poly (acrylic acid)–cysteine conjugates with regard to their molecular mass, *Eur. J. Pharm. Sci.*, *18*, 89–96.

[13] Senel, S., Kremer, M., Kas, S., Wertz, P. W., Hincal, A. A. & Squier, C. A. (2000). Enhancing effect of chitosan on peptide drug delivery across buccal mucosa, *J. Biomater.*, *21*, 2067–2071.

[14] Felt, O., Furrer, P., Mayer, J. M., Plazonnet, B., Buri, P. & Gurny, R. (1999). Topical use of chitosan in ophthalmology: Tolerance assessment and evaluation of precorneal retention, *Int. J. Pharm.*, *180*, 185–193.

[15] Felt, O., Buri, P. & Gurny, R. (1998). Chitosan: A unique polysaccharide for drug delivery, *J. Drug Dev. Ind. Pharm.*, *24*, 979–993.

[16] Kast, C. E., Frick, W., Losert, U. & Andreas, B. S. (2003). Chitosan–thioglycolic acid conjugate: A new scaffold material for tissue engineering, *Int. J. Pharm.*, *256*, 183–186.

[17] Shah, R. B., Ahsan, F. & Khan, M. A. (2002). Oral delivery of proteins: Progress and prognostication, *Crit. Rev. Ther. Drug Carrier Syst.*, *19*, 135–169.

[18] Coppi, G., Lannuccelli, V., Leo, E., Bernabei, M. T. & Cameroni, R. (2001). Chitosan–alginate microparticles as a protein carrier, *J. Drug Dev. Ind. Pharm.*, *27*, 393–400.

[19] Guggi, D., Kast, C. E. & Andreas, B. S. (2003). *In vivo* evaluation of an oral calcitonin delivery system for rats based on a thiolated chitosan matrix, *J. Pharm. Res.*, *20* (12), 1989–1994.

[20] Bromberg, L. E. (2001). Enhanced nasal retention of hydrophobically modified polyelectrolytes, *J. Pharm. Pharm.*, *53*, 109–114.

*Chapter 4*

# MUCOADHESIVE RECTAL DRUG DELIVERY SYSTEMS

The rectal route of administration (ROA) is a way of administering drugs into the rectum to be absorbed by rectum's blood vessels and into the body's circulatory systems which can then distribute the drug to the body's organs and various systems where the drug elicits its effects. A drug that is administered rectally will in general have a faster onset, higher bioavailability, shorter peak, and shorter duration than the oral route.

Another advantage of administering a drug rectally is that it tends to produce less nausea compared to the oral route. It also prevents any amount of the drug from being lost due to emsis since the drug is in the rectum, not the stomach, and the contents of the rectum are not lost when there is emesis. In addition, the rectal route bypasses around two-thirds of the first-pass metabolism as the rectum's venous drainage is two-thirds systemic and one-third portal. This means the drug will reach the circulatory system with significantly less alteration and in greater concentrations.

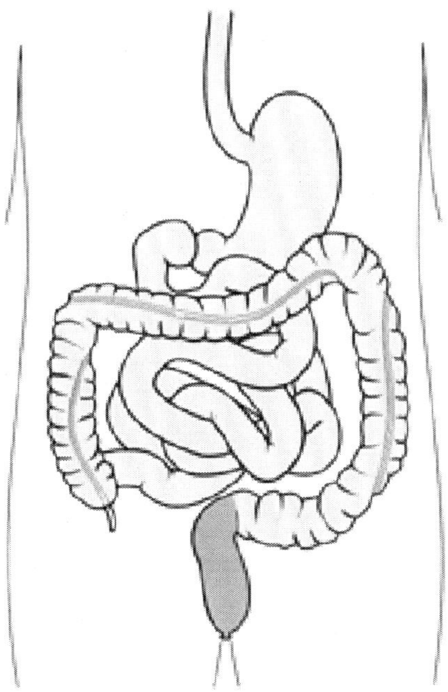

Figure 1. The diagram shows the position of rectum in the body.

## Rectal Transmucosal Administration

Medications may be administered by the rectal mucosal route for systemic effects if other more preferable routes are not available for the treatment of nausea and vomiting, sedation, control of seizures, analgesia, or antipyresis. Rectal administration provides rapid absorption of many drugs and may be an easy alternative to the intravenous route, having the advantage of being relatively painless, and usually no more threatening to children than taking a temperature. However, rectal administration of drugs should be avoided in immunosuppressed patients for whom even minimal trauma could lead to an abscess formation. The most important concern for the practitioner is the irregular uptake; clinically important patient-to-patient variability exists. The absorption of the drug may be delayed or prolonged, or uptake may be almost as rapid as if an intravenous bolus were administered, which may cause adverse cardiovascular or central nervous system effects. One reported death after rectal administration of multiple doses of morphine underscores the importance of being aware of this factor.

The rate of rectal transmucosal absorption is affected by the following factors:

- Formulation (time to liquefaction of suppositories)
- Volume of liquid
- Concentration of drug
- Length of rectal catheter (site of drug delivery)
- Presence of stool in the rectal vault
- pH of the rectal contents
- Rectal retention of drug(s) administered
- Differences in venous drainage within the rectosigmoid region

Despite the limitations associated with drug absorption in the rectum, many drugs usually administered by the intravenous and orogastric routes have also been administered rectally. Sedatives commonly administered by this route include midazolam, diazepam, and ketamine. In children, the rectal route is convenient for the administration of benzodiazepines to treat status epilepticus because an intravenous line is not required. The rectal dose generally must be higher than the dose administered intravenously or orally. The extent of the increase depends on the factors that affect absorption (listed earlier). The most important considerations are the slow onset of effect (minutes) and the prolonged duration of effect (hours). The peak blood levels vary considerably from patient to patient. The potential for rapid and almost complete absorption has serious implications when drugs with cardiac or pulmonary depressant effects are administered. Practitioners must be prepared to monitor the patient after drug administration and to manage an emergency should it occur; equipment suited to the size of the patient is required. The patient also may expel an unmeasurable amount of the drug, which makes it difficult for the practitioner to decide how much more of the drug to administer. It has been accepted for years medications can be administered in the rectum. Antiemetics are often prescribed by this route, and anesthesiologists have administered preinduction medications rectally for quite some time. More recently, a rectal gel became available for the treatment of status epilepticus, and many parents of children with seizure disorders have this agent in their homes for emergency use. Although rectal administration is convenient in children who are vomiting or uncooperative, as with every means of drug delivery, there are important considerations. One of the key issues with rectal drug administration is the venous drainage of the rectum. The veins in the upper portion of the rectum (the superior rectal veins) are part of the portal circulation, and agents administered in this portion of the rectum are often transported directly to the liver. If the agent administered this way is subject to a significant first-pass effect, it will be metabolized before a clinical effect is seen, and may be ineffective. On the other hand, the middle and inferior rectal veins drain into the systemic circulation, and agents absorbed in this way are avoiding the first-pass effect. Additionally, there appears to be significant patient-to-patient variability in the rate of absorption. In some children, drug levels peak quickly, almost as if an intravenous bolus had been given, while in others absorption is significantly delayed. Of course, the presence of stool in the rectal vault may interfere with absorption. Other factors help determine the effect of rectally administered medications. Liquid medications can be expelled by the rectum, but if they are retained, they are absorbed rather quickly. Suppositories must liquefy before they can be absorbed, meaning they are absorbed more slowly. The volume of the liquid administered and the concentration of the drug are also important factors. Patients were equally well sedated by 0.25 cc/kg of a 10% methohexital solution and by 0.75 cc/kg of a 2% solution, but peak drug levels were sustained longer in the children who received the 2% solution. This implies that the larger volume was absorbed gradually, leading to a more protracted period of sedation. The rectal contents also can affect drug delivery. If there is stool in the vault or if rectal pH does not favor absorption of the medication in question, then absorption will be slowed. While all of these factors influence the effectiveness of a rectally administered drug, in fact, many medications including benzodiazepines, barbiturates, ketamine, and antiemetics are successfully administered rectally hundreds of times each day. Nonetheless, their impact on an individual patient cannot be predicted with certainty. Peak blood levels achieved by identical per-kilogram doses of medication may be very different. This is the primary problem with rectal administration, especially for emergency physicians. If Johnny needs sedation for an outpatient MRI and his rectally administered sedative does not achieve the desired effect, he will have to come back another day. In the emergency department, the procedure often cannot be delayed or postponed. If Johnny is in status epilepticus and lacks intravenous access, we obviously cannot bring him back tomorrow for another try. Rectal medication has a somewhat circumscribed role in the ED because of several limitations. Unreliable rectal absorption also can be potentially dangerous. To be certain of achieving effective serum drug levels, it is often necessary to administer a somewhat higher per-kilogram dose of medication than might be administered intravenously or orally. Suppose that we have given such a dose of medication to little Johnny to sedate him for his procedure, but after waiting a sufficient period of time, we conclude that the rectal medication is not effective. So we bite the bullet, start an IV, and then administer an intravenous sedative. Unknown to us, however, the rectal medication that we gave Johnny is slowly being absorbed, and at some point in the procedure, we notice that Johnny isn't breathing well. Given these limitations, rectal medication has a somewhat circumscribed role in the emergency department. Rectally administered anticonvulsants should be considered for the management of status epilepticus when intravenous access is not available. Rare patients might be appropriate candidates for rectally administered sedatives and antiemetics. In nearly 20 years of pediatric emergency medicine, however, I have only ordered a rectal medication other than acetaminophen a handful of times. Medications may be administered by the rectal mucosal route for systemic effects if other more preferable routes are not available for the treatment of nausea and vomiting, sedation, control of seizures, analgesia, or antipyresis. Rectal administration provides rapid absorption of many drugs and may be an easy alternative to the intravenous route, having the advantage of being relatively painless, and usually no more threatening to children than taking a temperature. However, rectal administration of drugs should be avoided in immunosuppressed patients in whom even minimal trauma could lead to formation of an abscess. The most important concern for a practitioner is the irregular uptake; clinically important patient-to-patient variability exists. The absorption of the drug may be delayed or prolonged, or uptake may be almost as rapid as if an intravenous bolus were administered, which may cause adverse cardiovascular or central nervous system effects. One reported death after rectal administration of multiple doses of morphine underscores the importance of being aware of this factor. Anatomical differences in hemorrhoidal venous drainage of the rectum may substantially influence the systemic drug level achieved. Drugs administered high in the rectum are usually carried directly to the liver and, thus, are subject to metabolism. Drugs administered low in the rectum are delivered systemically by the inferior and middle rectal before passing through the liver. Problems may occur with drugs that normally have a high hepatic extraction ratio. The clinical implications of rectal venous drainage for absorption and metabolism of most drugs are not well-defined. Diluent volume is also a determinant of rectal drug uptake, as demonstrated with methohexital administered rectally for preprocedure sedation. Equivalent deep sedation was achieved with 25mg/kg of a 10% solution and with 15 mg/kg of a 2% solution. Peak blood levels of the drug, however, were significantly higher for a longer time in the children treated with the 2% solution. Rectal pH may also influence drug uptake by altering the amount of drug that is ionized. The greater lipid solubility of nonionized drugs enhances their movement across biological membranes. The pH of the rectal vault in children ranges from 7.2 to 12.2. This range favors absorption of barbiturates that will remain in a nonionized state because their $pK_a$ is near the physiological range (~7.6). Despite the limitations associated with drug absorption in the rectum, many drugs usually administered by the intravenous and orogastric routes have also been administered rectally. The rate of rectal transmucosal absorption is affected by the important factors: such as, formulation, volume of liquid, concentration of drug, length of rectal catheter, presence of stool in the rectal vault, pH of the rectal contents, rectal retention of drugs administered, and differences in venous drainage within the rectosigmoid region. Oral and rectal misoprostol have been used for routine management of the third stage of labor. The main side effects have been shivering and pyrexia, which are dose-dependent. Physiological studies have also shown a more rapid onset of uterine contractions following syntometrine than misoprostol after delivery. Misoprostol has been widely recommended for the prevention of postpartum hemorrhage when other methods are not available. Rectal route of administration have been proved beneficial in approaching this problem. The development of a rectal formulation of Mebeverine hydrochloride with prolong drug action is expected to be appropriate for the control of the irritable bowel syndrome. The use of rectal route for sustaining drug action has been successfully reported with a number of drugs such as nifedipine, morphine, and metoclopramide HCl. However, conventional suppositories tend to be of solid dosage form that melt or soften in the rectum, may give an alien feeling, or discomfort, with the potential of refusal from the patients; possibly lowering patient compliance which represents a problem especially when treating a chronic disease such as the irritable bowel syndrome. To solve the problem of these conventional solid dosage forms, there have been several attempts to develop a rectal dosage form which exists as liquid at room temperature but can gel physiologically. The formed gel is expected to be softer than the conventional suppository. Poloxamers are surfactant copolymers of poly(oxyethylene)-poly(oxypropylene)-poly(oxyethylene) exhibiting such thermo reversible property. By modulating the gelation temperature of different poloxamer solutions, liquid bases for rectal use can be formulated which form a gel in the rectum at body temperature with suitable gel strength; so, as not to leak out from the anus after administration. In addition, the incorporation of mucoadhesives in these bases would reinforce the drug sustained action, and may help the formed gel to adhere to the lower part of the rectum offering further protection from the first pass effect. Hence, improving its bioavailability. Mucoadhesive suppositories of propanolol HCl and lidocaine were previously formulated to achieve such protection from the first effect. Organ culture (OC) is the most common corneal storage method in Europe, used for more than two-thirds of donor corneas. The extended storage time of 4 to 5 weeks facilitates graft quality and safety controls in eye banks and operation scheduling. However, all OC media have the drawback to often trigger rapid stroma swelling, resulting in doubled corneal thickness, reduced corneal transparency, and increased stroma and folds in Descemet's membrane. To facilitate suturing and accelerate postoperative visual recovery, the cornea must be thinned before surgery by immersion in the same OC medium supplemented with a water-soluble macromolecule that produces colloid osmotic pressure to extract excess water accumulated in the corneal stroma. Dextrans are polyglucose biopolymers of 1000 to 2000 kDa that have a high affinity for water because of their abundant hydrophilic hydroxyl groups. Technical (T) dextrans are high-purity dextran fractions with selected average molecular weights. Dextran T250 and T500 (500 kDa) were first introduced by Sperling in 1979, and T500 is to date the only macromolecule routinely used for corneal deswelling. Initially thought usable throughout corneal storage, dextran was in practice rapidly restricted to the final 48 to 72 hours of storage. However, this was initially described as an innocuous effect for dextran-containing media on corneal endothelial cells (ECs) and has been questioned. An average of 8.4% EC loss was reported after 1 to 4 days of deswelling in an OC medium supplemented with 5% dextran T500. A 15% cell loss after 2 days of deswelling in commercial media containing 5% dextran T500, a result consistent as reported previously. Daily EC loss in dextran-containing media consequently appears much higher than average daily EC loss of 0.9% in dextran-free media used in the first phase of OC. Two other macromolecules have more recently been evaluated for corneal deswelling to replace dextran in OC. Hydroxyethyl starch (HES) 130 and 450, vegetable-origin polysaccharides initially used for plasma expansion, have been assessed *ex vivo* in pig and human corneas. HES130 has been assessed as a supplement throughout OC. It restricts water accumulation during long-term storage, making pregraft deswelling unnecessary. Its effect on ECs, however, has not been directly compared with dextran. No clinical application has yet been reported. The second macromolecule assessed belongs to the poloxamer family. Poloxamers are synthetic nonionic triblock copolymers comprising a central hydrophobic polyoxypropylene molecule flanked on each side by a hydrophilic chain of polyoxyethylene. They were introduced in the 1950s by BASF (Ludwigshafen, Germany). Their molecular weights range from 2,000 to 18,000. The physicochemical characteristics of different poloxamers depend on the proportion and the size of both subunits, as well as on their concentration in the solution. In the pharmaceutical field, poloxamers, per se, have therapeutic properties used to treat vaso-occlusive crisis of sickle cell disease, but they are mostly used simply as drug excipients. In ophthalmology, poloxamers have been proposed for inclusion in the formulation of certain eye drops to improve their pharmacokinetic properties. They are also used as an antiadhesive agent against bacteria in contact lens-cleaning solutions. Poloxamer hydrogel has been assessed as a well-tolerated injectable intraocular lens in the rabbit. Findings in our study of an animal compound-free (ACF) medium combined with poloxamer 188 for corneal deswelling suggest that it is superior to the standard fetal calf serum (FCS) medium containing dextran T500 in preserving EC viability during deswelling. A clinical trial comparing these media is ongoing. Nevertheless, the aforementioned study was not designed solely to assess the effect of poloxamer independently of ACF. Consequently, the purpose of the present study was to assess the innovative use of five poloxamers for human corneal deswelling in OC by comparison with dextran T500, with a standard OC medium containing 2% FCS. Compared with dextran, four of the five poloxamers assessed were found superior in reducing EC loss and had similar efficacy in corneal deswelling and a comparable effect on corneal ultrastructure.

Rectally administered preparations of 5-aminosalicylic acid should be the preferred treatment for mildly to moderately active left-sided or distal ulcerative colitis. However, sulphasalazine and oral derivatives of 5-aminosalicylic acid, which have played a major role in the treatment of Crohn's disease and ulcerative colitis for several decades, remain the most commonly used therapies for even distal ulcerative colitis. Several dosage forms of 5-aminosalicylic acid have been evaluated in clinical trials assessing rectally administered treatment for active distal ulcerative colitis. These preparations include suppositories, foams, and liquid enemas, which differ with respect to their physicochemical properties and potential for proximal spread. The efficacy of most 5-aminosalicylic acid therapies is dependent upon direct contact with the inflamed mucosa. Thus, for the proximal disease margin, the patient's willingness to comply, and the ability to retain a rectal treatment will determine which dosage form is the most appropriate. It is important to note that most studies that have evaluated the distribution of rectal therapies have employed techniques that would be difficult to replicate in clinical practice. For example, some have required prolonged fasting or supine positioning for up to 4 hours without turning. Another important factor in the ability to retain rectal therapies is the disease severity. Despite these drawbacks, improvements in the mode of administration are likely to enhance patient compliance and clinical effectiveness of rectal therapy. A recent meta-analysis identified that patient selection, failure to stratify by disease extent, and variability in duration of therapy represented significant flaws in the research methods of trials of rectal therapy. The variable dosing, volumes, and formulations have left some important questions unanswered. The dose, volume, and delivery vehicles of the rectal 5-aminosalicylic acid preparations varied substantially among the seven placebo-controlled trials eligible to be pooled for meta-analysis. Three studies assessed 5-aminosalicylic acid suppositories, with the daily dose ranging from 0.5 g to 1.5 g, whereas the remaining four trials evaluated liquid 5-aminosalicylic acid enemas with daily doses of 1.0-4.0 g and volumes of 60-100 mL. Cylindrical hydrogels of hydroxyethyl methacrylate (HEMA) and ethylene glycol dimethacrylate (EGDMA) as cross-linking agent were prepared by radical polymerization at 70°C. After washing, they were soaked in an aqueous drug solution of antipyrine or theophylline. The *in vitro* drug release experiments were performed in 100 ml isotonic glucose at 37°C. Rectal administration of a hydrogel preparation containing antipyrine was performed in two subjects for 72 h. For the theophylline-containing hydrogel preparation, the rectal drug administration was performed in six volunteers for 24 h. Plasma and saliva samples were taken regularly and the *in vivo* drug release was determined by means of a deconvolution procedure. *In vitro* 1.12g antipyrine had been released according to a matrix-type profile for 72 h, whereas it was calculated that this was 1.13 and 1.09 g *in vivo* in the two subjects. The release profile *in vivo* was very similar to the *in vitro*. The theophylline hydrogel preparation released *in vitro* a total of 288 ± 6 *mg* of drug in 24 h and *in vivo* this amount was calculated to be 288 ± 11 *mg* (*mean* ± *s.d.*). Near-constant plasma theophylline concentrations were obtained after administering the hydrogel preparation. In all six subjects, the cumulative drug profile was in almost perfect agreement with that observed *in vitro*. Hydrogels offer interesting perspectives as rate-controlled rectal drug delivery systems because of the predictable release profile *in vivo* on the basis of observations in a simple *in vitro* model. *In situ* gels also possess a potential application for drug delivery by rectal and vaginal route. Administration of indomethacin loaded xyloglucan based systems to rabbits indicated broad drug absorption peak and a longer drug residence time as compared to that resulting after the administration of commercial suppository. In addition, a significant reduction of drug $C_{max}$ was observed after administration of *in situ* polymeric system thus indicating the avoidance of adverse effects of indomethacin on nervous system. For a better therapeutic efficacy and patient compliance, a mucoadhesive, thermo sensitive, prolonged release vaginal gel incorporating clotrimazole-β-cyclodextrin complex was formulated for the treatment of vaginitis. Pluronic F-127 was used as an *in situ* gel forming polymer together with mucoadhesive polymers such as Carbopol 934 and hydroxylpropylmethylcellulose to ensure long residence time at the application site. Controlled release of the drug was achieved *in vitro* indicating antimycotic efficacy of developed formulation for a longer period of time. Insulin is a polypeptide drug and it is degraded by gastrointestinal enzymes, therefore, it cannot be used via oral route readily. There are only parenteral forms available in the market. Chitosan gel (CH-gel) was used as a carrier; the penetration enhancing effect of sodium taurocholate and dimethyl-β-cyclodextrin (DM-βCD) was also investigated. CH-gel provided longer insulin release. The maximum decreasing effect on blood glucose level was observed with insulin–CH-gel containing 5% DM-βCD. The insulin may penetrate well through the rectal and vaginal mucosa from the CH-gel. DM-βCD was also found to be a useful agent to enhance the penetration of insulin through rectal and vaginal membranes. Recent studies have shown that rectal mucosa can be used as a potential site for delivering drugs which have a bitter and nauseous taste, have a high first-pass metabolism and degrade in the gastrointestinal pH. It is an ideal route to deliver drugs to the unconscious patients, children, and infants. However, rectal mucosa offers a very limited area for drug absorption resulting in an erratic release of drugs.

**Table 1. CDs in rectal drug delivery system**

| Improvement | Drug |
|---|---|
| Enhanced drug release | Ethyl-4-biphenylyl acetate (EBA), Insulin |
| Enhanced rectal absorption | Morphine, Bis (acetate) amine, Dichloro(cyclohexyl amine) platinum(IV) |
| Enhanced chemical stability | Human granulocyte colony stimulating factor |
| Reduced rectal irritation | Prednisolone |
| Enhanced bioavailability | Morphine, Human chorionic gonadotrophin |

To overcome these problems, a number of excipients have been used, and amongst them CDs have been found to be quite useful. Table 1 shows CDs in rectal drug delivery system. To be used as excipients in rectal drug delivery system, CDs should have the following characteristics.

## REFERENCES


[1] De Blaey, C. J. & Tukker, J. J. (1988). Suppositories and pessaries. In: M.E. Aulton (ed.), *Pharmaceutics: The Science of Dosage Form Design*, 3RD edit., Leicester,UK, Churchill Livingstone, 412-414.

[2] Rossi, S., Sandri, G., Ferrari, F., Bonferoni, M. C. & Caramells, C. (2003). Development of films and matrices based on chitosan and polyacrylic acid for vaginal delivery of acyclovir. *STP Pharma Sci.*, *13*, 183-198.

[3] Illum, L., Wright, J. & Davis, S. S. (1989). Targeting of microspheres to sites of inflammation. *Int. J. Pharm.*, *52*, 221-224.

[4] Davis, S. S. & Illum, L. (1983). The targeting of drugs using polymeric microspheres. In: S.S. Davis and L. Illum (eds.), *Polymeric Drug Delivery System*, 2ND edit., Nothingham, UK, University of Nothingham Press, 1-113.

[5] Coombes, A., Tasker, S., Lindblal, M., Holmgreen, K. & Hoste, V. (1997). Biodegradable polymeric microparticles for drug delivery and vaccine formulation: The surface attachment of hydrophilic species using the concept of poly(ethylene)glycol anchoring segments. *Biomaterials*, *18*, 1153-1161.

[6] Ida, G., Monica, C., Annalia, A., Bice, C. & Luisa, M. (1998). Influence of glutaraldehyde on drug release and mucoadhesive properties of chitosan microspheres. *Carbohyd Polym. Sci.*, *42*, 81-88.

[7] Albertsson, A. C., Carlfors, J. & Sturesson, C. (1995). Preparation and characterization of poly (adipic anhydride) microspheres for ocular drug delivery. *J. Appl. Polym. Sci.*, *62*, 695-705.

[8] Kedziericz, F., Lombry, C., Rios, R. & Hoffman, P. M. (1999). Effect of the formulation on the *in vitro* release of propranolol from gellan beads. *Int. J. Pharm.*, *178*, 129-136.

[9] Santucci, E., Alhaique, F., Carafa, M. & Coviello, T. (1996). Gellan for the formulation of sustained delivery beads. *J. Control Rel.*, *42*, 157-164.

[10] Gupta, K. C. & Ravi-Kumar, M. N. V. (2000a). Drug release behavior of beads and micro granules of chitosan. *Biomaterials*, *21*, 1115-1119.

[11] Gupta, K. C. & Ravi-Kumar, M. N. V. (2000b). Semi-interpenetrating polymer network beads of cross-linked chitosanglycine for controlled release of chlorpheniramine maleate. *J. Appl. Polym. Sci.*, *76*, 672-683.

[12] Wakiyama, N. J. K. & Nakano, M. (1982). Preparation and evaluation of *in vitro* and *in vivo* polylactic acid microspheres containing dibucaine. *Chem. Pharm. Bull*, *30*, 3719-3727.

[13] Alpar, H. O., Field, W. N., Hyde, R. & Lewis, D. A. (1989). The transport of microspheres from the gastrointestinal tract to inflammatory air pouches in the rat. *J. Pharm. Pharmacol.*, *41*, 194-196.

[14] Yin, Y., Xu, M., Chen, X. & Yao, K. O. (1996). Drug release behavior of chitosan/gelatin network polymer microspheres. *Chinese Sci. Bull*, *41*, 1266-1270.

[15] Mortazari, S. A., Carpenter, B. G. & Smart, J. D. (1993). A comparative study on the role played by mucus glycoproteins in the rheological behavior of the mucoadhesive/mucosal interface. *Int. J. Pharm.*, *94*, 195-207.

[16] Stone, S. H. (1954). Methods for obtaining venous blood from the orbital sinus of a rat or mouse. *Science*, *119*, 100-102.

[17] Riley, V. (1950). Adaptation of orbital bleeding technique to rapid serial blood studies. *Proc. Soc. Exper. Biol.Medicine*, *104*, 751-755.

[18] Tietz, N. W. (1970). Preparation of protein-free filtrate. In: N.W. Tietz (ed.), *Fundamentals of Clinical Chemistry*, 2ND edit., Philadelphia, USA, W. B. Saunders Company, 32-45.

[19] Higuchi, T. (1963). Mechanism of sustained action medication: Theoretical analysis of rate of release of solid drugs dispersed in solid matrices. *J. Pharm. Sci.*, *52*, 1145-1149.

[20] Peppas, N. A.(1985). Analysis of Fickian and non-Fickian drug release from polymers. *Pharm. Acta Helv.*, *60*, 110-111.

[21] Philip, J. C., Karrar, A. K., Dale, L. M. & Jomjai, S. (1999). Development and evaluation of a multiple-unit oral sustained release dosage form for S(+)-ibuprofen: Preparation and release kinetics. *Int. J. Pharm.*, *197*, 73-84.

*Chapter 5*

# Mucoadhesive Buccul Delivery Systems

Bioadhesive polymers are defined as those polymers that can adhere onto a biological substrate. The term mucoadhesion is applied when the substrate is a mucosal tissue. Polymers which can adhere to either hard or soft tissue have been used for many years in surgery and dentistry. Diverse classes of polymers have been investigated for their potential use as mucoadhesives. These include synthetic polymers such as monomeric cyanoacrylate, polyacrylic acid, hydroxypropyl methylcellulose, and poly methacrylate derivatives. There are also naturally occurring polymers, such as hyaluronic acid and chitosan. Other synthetic polymers such as polyurethanes, epoxy resins, polystyrene, and natural-product cement have also been extensively investigated.

The dosage forms designed for buccal administration should not cause irritation and should be small and flexible enough to be accepted by the patient. These requirements can be met by using hydrogels. Hydrogels are hydrophilic matrices that are capable of swelling when placed in aqueous media. Normally it can only absorb water. When drugs are loaded into these hydrogels, as water is absorbed into the matrix, chain relaxation occurs and drug molecules are released through the spaces or channels within the hydrogel network. In a broader meaning of the term, hydrogels would also include water-soluble matrices that are capable of swelling in aqueous media, as to include natural gums and cellulose derivatives. These 'pseudo hydrogels' swell infinitely and the component molecules dissolve from the surface of the matrix. Drug release would then occur through the spaces or channels within the network, as well as through the dissolution and/or the distintegration of the matrix.

A novel delivery system has been developed for testosterone replacement. This formulation describes a testosterone-containing buccal mucoadhesive system, which has been shown in preliminary studies to replace testosterone at physiological levels when used twice daily. Mean serum testosterone concentration during the 24-hour treatment period was 18.74±5.90 nmol/L in the buccal system group and 12.5±5.55 nmol/L in the transdermal patch group. In the buccal system group, 97% of patients had average steady-state testosterone concentrations within the physiological range compared with 56% in the transdermal patch group. Testosterone concentrations were within the physiological range for a greater portion of the 24-hour treatment period in the buccal system group than in the transdermal patch group.

The buccal mucosa lines the inner cheek, and buccal formulations are placed in the mouth between the upper gingivate and cheek to treat local and systemic conditions. The buccal route provides a potential route for typically large, hydrophilic and unstable proteins, oligonucleotides, and polysaccharides, as well as conventional small drug molecules. The oral cavity has been used as a site for local and systemic drug delivery. Advantages of drug delivery via the buccal lining include: 1) Bypass of gastroinintestinal tract and hepatic portal system, increasing the bioavailability of orally administered drugs that otherwise undergo hepatic first-pass metabolism. In addition, the drug is protected from degradation due to pH and digestive enzymes of the middle gastrointestinal tract. 2) Improved patient compliance due to elimination of associated pain with injections, administration of drugs in unconscious of incapacitated patients, convenience of administration as compared to injections or oral medications. 3) Sustained drug delivery. 4) A relatively rapid onset of action can be achieved relative to the oral route, and the formulation can be removed if therapy is required to be discontinued. 5) Increased ease of drug administration. 6) Though less permeable than the sublingual area, the buccal mucosa is well vascularized, and drugs can be rapidly absorbed into the venous system underneath the oral mucosa. 7) Transmucosal delivery occurs is less variable between patients, resulting in lower subject variability as compared to transdermal patches. 8) The large contact surface of the oral cavity contributes to rapid and extensive drug absorption. The rate of absorption of hydrophilic compounds is a function of the molecular size. Smaller molecules (75-100 Da) generally exhibit rapid transport across the mucosa, with permeability decreasing as molecular size increases. For hydrophilic macromolecules such as peptides, absorption enhancers have been used to successfully alter the permeability of the buccal epithelium, causing this route to be more suitable for the delivery of larger molecules. Only the nonionized forms of molecules have the ability to cross-lipoidal membranes in significant amounts. A compound that has the ability to be more lipid-soluble than the higher the permeability can be. The permeability of these compounds is a direct functions of their oil-water partition coefficients. The partition coefficient is a useful tool to determine the absorption potential of a drug. In general, increasing a drug's polarity by ionization or the addition of hydroxyl, carboxyl, or amino groups, will increase the water solubility of any particular drug and cause a decrease in the lipid-water partition coefficient. Conversely, decreasing the polarity of a drug (e.g., adding methyl or methylene groups) results in an increased partition coefficient and decreased water solubility. The partition coefficient is also affected by pH at the site of drug absorption. When increasing pH, the partition coefficient of acidic drugs decreases, while that of basic drugs increases. The partition coefficient is also an important indicator of drug storage in fat deposits. Obese individuals can store large amounts of lipid-soluble drug in their fat. These drugs are dissolved in the lipid and are a reservoir of slow release from these fat deposits. The ionization of a drug is directly related to both its pKa and pH at the mucosal surface. Only the nonionized form of many weak acids and weak bases exhibit appreciable lipid solubility, and thus the ability to cross lipoidal membranes. As a result, maximal absorption of these compounds has been shown to occur at the pH at which they are unionized, with absorbability diminishing as ionization increases. In short, one can say that the lipid solubility of drugs is an important factor in Transmucosal Drug Delivery system. Along with lipid solubility, drugs selected for Transmucosal Drug Delivery system must have physiochemical properties, including size and pKa that facilitate drug movement through the mucosa at a rate capable of producing therapeutic blood concentrations. The drug must resist, or be protected by salivary and tissue enzymes that could cause inactivation. Additionally, the drug and adhesive materials must not damage the teeth, oral cavity, or surrounding tissues (e.g., by keratinolysis, discoloration, and irritation). There are two possible routes of drug absorption through the squamous stratified epithelium of the oral mucosa: 1) Transcellular (intracellular, passing through the cell), and 2) Paracellular (intercellular, passing around the cell). Permeation across the buccal mucosa has been reported to be mainly by the paracellular route through the intercellular lipids produced by membrane-coating granules. Although passive diffusion is the main mechanism of drug absorption, specialized transport mechanisms have been reported to exist in other oral mucosa (that of the tongue) for a few drugs and nutrients; glucose and cefadroxil were shown to be absorbed in this way. The buccal mucosa is a potential site for the controlled delivery of hydrophilic macromolecular therapeutic agents (biopharmaceuticals) such as peptides, oligonucleotides, and polysaccharides. However, these high molecular weight drugs usually have low permeability leading to a low bioavailability, and absorption enhancers may be required to overcome this. The buccal mucosa also contains proteases that may degrade peptide-based drugs. In addition, the salivary enzymes may also reduce stability. Disease states where the mucosa is damaged would also be expected to increase permeability. This would be particularly true in conditions that result in erosion of the mucosa such as lichen planus, pemphigus, viral infections, and allergic reactions. Absorption enhancers have demonstrated their effectiveness in delivering high molecular weight compounds, such as peptides, that generally exhibit low buccal absorption rates. These may act by a number of mechanisms, such as increasing the fluidity of the cell membrane, extracting inter/intracellular lipids, altering cellular proteins or altering surface mucin. The most common absorption enhancers are azone, fatty acids, bile salts, and surfactants; such as sodium dodecyl sulfate. Solutions/gels of chitosan were also found to promote the transport of mannitol and fluorescent-labeled dextrans across a tissue culture model of the buccal epithelium while Glyceryl monooleates were reported to enhance peptide absorption by a co-transport mechanism. Formulations that produce local damage at the site of application, such as ulceration of the mucosa, would preclude their widespread usage as a result of the associated pain and discomfort. This is particularly important in buccal drug delivery where the formulation is in contact with the mucosa for extended periods. Toxic effects can arise from the drug itself, the bioadhesive or from other components of the formulation. For example, carbomers have been reported to produce mucosal irritation believed to result from a localized low pH, whereas lectins have been shown to be cytotoxic. They are two permeation pathways for passive drug transport across the oral mucosa: paracellular and transcellular routes. Permeants can use these two routes simultaneously, but one route is usually preferred over the other depending on the physicochemical properties of the diffusant. Since the intercellular spaces and cytoplasm are hydrophilic in character, lipophilic compounds would have low solubilities in this environment. The cell membrane, however, is rather lipophilic in nature and hydrophilic solutes will have difficulty permeating through the cell membrane due to a low partition coefficient. Therefore, the intercellular spaces pose as the major barrier to permeation of lipophilic compounds and the cell membrane acts as the major transport barrier for hydrophilic compounds. Since the oral epithelium is stratified, solute permeation may involve a combination of these two routes. The route that predominates, however, is generally the one that provides the least amount of hindrance to passage. Even though the sublingual mucosa is relatively more permeable than the buccal mucosa, it is not suitable for an oral transmucosal delivery system. The sublingual region lacks an expanse of smooth muscle or immobile mucosa and is constantly washed by a considerable amount of saliva making it difficult for device placement. Because of the high permeability and the rich blood supply, the sublingual route is capable of producing a rapid onset of action making it appropriate for drugs with short delivery period requirements with infrequent dosing regimen. Due to two important differences between the sublingual mucosa and the buccal mucosa, the latter is a more preferred route for systemic transmucosal drug delivery. The first difference being in the permeability characteristics of the region, where the buccal mucosa is less permeable and is thus not able to give a rapid onset of absorption (*i.e.*, more suitable for a sustained release formulation). The second difference is that the buccal mucosa has an expanse of smooth muscle and relatively immobile mucosa which makes it a more desirable region for retentive systems used for oral transmucosal drug delivery. Thus the buccal mucosa is more fitted for sustained delivery applications, delivery of less permeable molecules, and perhaps peptide drugs. As stated earlier in section I, there are three different categories of drug delivery within the oral cavity (*i.e.*, sublingual, buccal, and local drug delivery). Selecting one over another is mainly based on anatomical and permeability differences that exist among the various oral mucosal sites. The sublingual mucosa is relatively permeable, giving rapid absorption and acceptable bioavailabilities of many drugs, and is convenient, accessible, and generally well accepted. The sublingual route is by far the most widely studied of these routes. Sublingual dosage forms are of two different designs, those composed of rapidly disintegrating tablets, and those consisting of soft gelatin capsules filled with liquid drug. Such systems create a very high drug concentration in the sublingual region before they are systemically absorbed across the mucosa. The buccal mucosa is considerably less permeable than the sublingual area, and is generally not able to provide the rapid absorption and good bioavailabilities seen with sublingual administration. Local delivery to tissues of the oral cavity has a number of applications, including the treatment of toothaches, periodontal disease, bacterial and fungal infections, aphthous and dental stomatitis, and in facilitating tooth movement with prostaglandins. Similar to any other mucosal membrane, the buccal mucosa as a site for drug delivery has limitations as well. One of the major disadvantages associated with buccal drug delivery is the low flux which results in low drug bioavailability. Various compounds have been investigated for their use as buccal penetration enhancers in order to increase the flux of drugs through the mucosa. Since the buccal epithelium is similar in structure to other stratified epithelia of the body, enhancers used to improve drug permeation in other absorptive mucosa have been shown to work in improving buccal drug penetration. Drugs investigated for buccal delivery using various permeation/absorption enhancers range in both molecular weight and physicochemical properties. Small molecules such as butyric acid and butanol, ionizable low molecular weight drugs such as acyclovir, propranolol, and salicylic acid, large molecular weight hydrophilic polymers such as dextrans, and a variety of peptides including octreotide, leutinizing hormone releasing hormone (LHRH), insulin, and a-interferon have all been studied. A series of studies on buccal permeation of buserelin and fluorescein isothiocyanate (FITC) labeled dextrans reported the enhancing effects of di- and tri-hydroxy bile salts on buccal penetration. Their results showed that in the presence of the bile salts, the permeability of porcine buccal mucosa to FITC increased by a 100-200 fold compared to FITC alone. The mechanism of penetration enhancement of FITC-labeled dextrans by sodium glycocholate (SGC) was shown to be concentration dependent. Below 10 mM SGC the buccal permeation was increased, when increasing the intercellular transport of 10 mM and higher concentrations caused the transcellular route to open up.

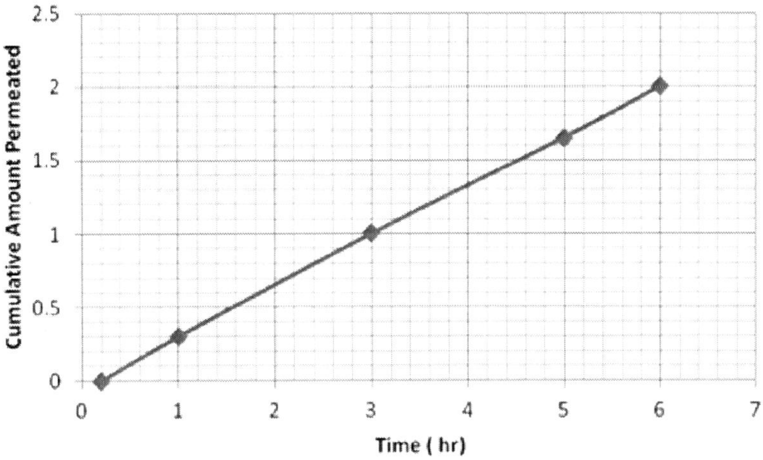


Gandhi and Robinson investigated the mechanisms of penetration enhancement in the transbuccal delivery of salicylic acid. They used sodium deoxycholate and sodium lauryl sulfate as penetration enhancers, both of which were found to increase the permeability of salicylic acid across rabbit buccal mucosa.

Their results also supported that the superficial layers and protein domain of the epithelium may be responsible for maintaining the barrier function of the buccal mucosa. A number of research groups have studied the feasibility of buccal mucosal delivery of insulin using various enhancers in different animal models for *in vivo* studies. Sodium glycocholate, sodium lauryl sulfate, sodium salicylate, sodium EDTA (ethylenediamine tetraacetic acid), and aprotinin on rat buccal mucosa noticed an increase in insulin bioavailability from about 0.7% (without enhancer) to 26-27% in the presence of sodium glycocholate (5% w/v) and sodium lauryl sulfate (5% w/v). Similar results were obtained using dog as the animal model for the *in vivo* studies, where sodium deoxycholate and sodium glycocholate yielded the highest enhancement of buccal insulin absorption. These studies have all demonstrated the feasibility of buccal delivery of a rather large molecular weight peptide drug such as insulin.

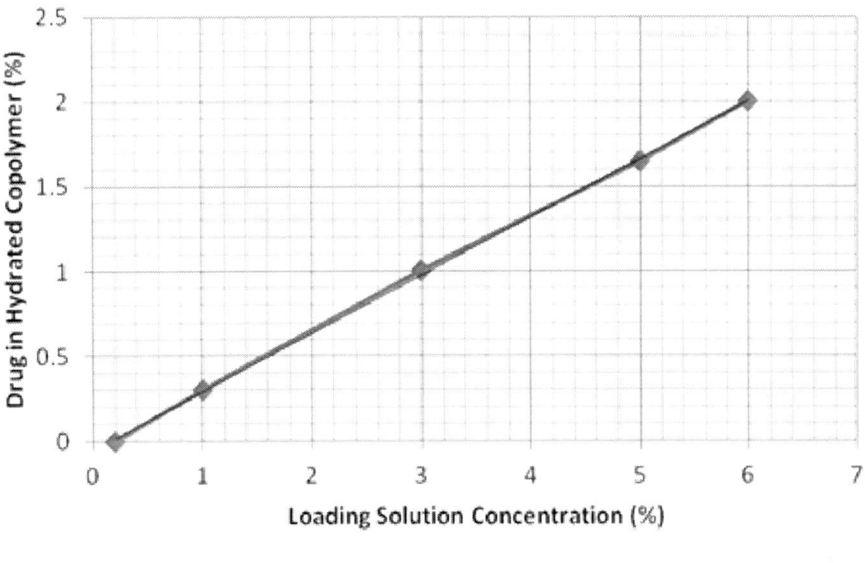

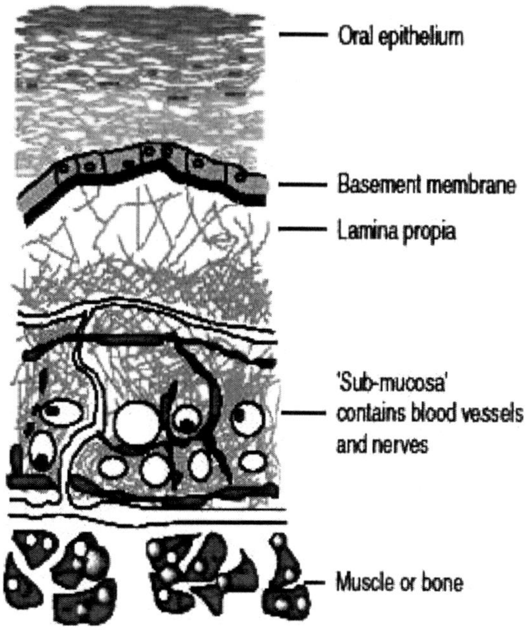


Figure 1. The layers of the Buccal drug delivery system.

Before a buccal drug delivery system can be formulated, buccal absorption/permeation studies must be conducted to determine the feasibility of this route of administration for the candidate drug. These studies involve methods that would examine *in vitro* and/or *in vivo* buccal permeation profile and absorption kinetics of the drug.

## *In Vivo* Methods

*In vivo* methods were first originated by Beckett and Triggs with the so-called buccal absorption test. In using this method the kinetics of the drug absorption were measured. The methodology involves the swirling of a 25 ml sample of the test solution for up to 15 minutes by human volunteers followed by the expulsion of the solution. The amount of drug remaining in the expelled volume is then determined in order to assess the amount of drug absorbed. The drawbacks of this method include salivary dilution of the drug, accidental swallowing of a portion of the sample solution, and the inability to localize the drug solution within a specific site (buccal, sublingual, or gingival) of the oral cavity. Various modifications of the buccal absorption test have been carried out correcting for salivary dilution and accidental swallowing, but these modifications also suffer from the inability of site localization. A feasible approach to achieve absorption site localization is to retain the drug on the buccal mucosa using a bioadhesive system. Pharmacokinetic parameters such as bioavailability can then be calculated from the plasma concentration vs. time profile. Other *in vivo* methods include those carried out using a small perfusion chamber attached to the upper lip of anesthetized dogs. The perfusion chamber is attached to the tissue by cyanoacrylate cement. The drug solution is circulated through the device for a predetermined period of time and sample fractions are then collected from the perfusion chamber (to determine the amount of drug remaining in the chamber) and blood samples are drawn after 0 and 30 minutes (to determine amount of drug absorbed across the mucosa).

## EXPERIMENTAL ANIMAL SPECIES

Aside from the specific methodology employed to study buccal drug absorption/ permeation characteristics, special attention is warranted to the choice of experimental animal species for such experiments. For *in vivo* investigations, many researchers have used small animals including rats and hamsters or permeability studies. However, such choices seriously limit the value of the data obtained since, unlike humans, most laboratory animals have an oral lining that is totally keratinized. The rat has a buccal mucosa with a very thick, keratinized surface layer. The rabbit is the only laboratory rodent that has non-keratinized mucosal lining similar to human tissue and has been extensively utilized in experimental studies. The difficulty in using rabbit oral mucosa, however, is the sudden transition to keratinized tissue at the mucosal margins making it hard to isolate the desired non-keratinized region. The oral mucosa of larger experimental animals that has been used for permeability and drug delivery studies include monkeys, dogs, and pigs. Due to the difficulties associated with maintenance of monkeys, they are not very practical models for buccal drug delivery applications. Instead, dogs are much easier to maintain and considerably less expensive than monkeys and their buccal mucosa is non-keratinized and has a close similarity to that of the human buccal mucosa. Pigs also have non-keratinized buccal mucosa similar to that of human and their inexpensive handling and maintenance costs make them an equally attractive animal model for buccal drug delivery studies. In fact, the oral mucosa of pigs resembles that of human more closely than any other animal in terms of structure and composition. However, when used during *in vivo* studies, pigs were not as ideal as dogs, due to their rapid growth which renders the animal handling rather difficult. Miniature breeds of pigs can be used but their high cost is a deterrent. For *in vitro* studies though, because of easy availability and low cost porcine tissue is more suited as compared to dog buccal tissue.

## *IN VITRO* METHODS

At the present time, most of the *in vitro* studies examining drug transport across buccal mucosa have used buccal tissues from animal models. Animals are sacrificed immediately before the start of an experiment. Buccal mucosa with underlying connective tissue is surgically removed from the oral cavity, the connective tissue is then carefully removed and the buccal mucosal membrane is isolated. The membranes are then placed and stored in ice-cold (4°C) buffers (usually Krebs buffer) until mounted between side-by-side diffusion cells for the *in vitro* permeation experiments. The most significant questions concerning the use of animal tissues as *in vitro* models in this manner are the viability and the integrity of the dissected tissue. How well the dissected tissue is preserved is an important issue which will directly affect the results and conclusion of the studies. To date, there are no standard means by which the viability or the integrity of the dissected tissue can be assessed. The scientists studied tissue viability by using ATP levels in rabbit buccal mucosa.

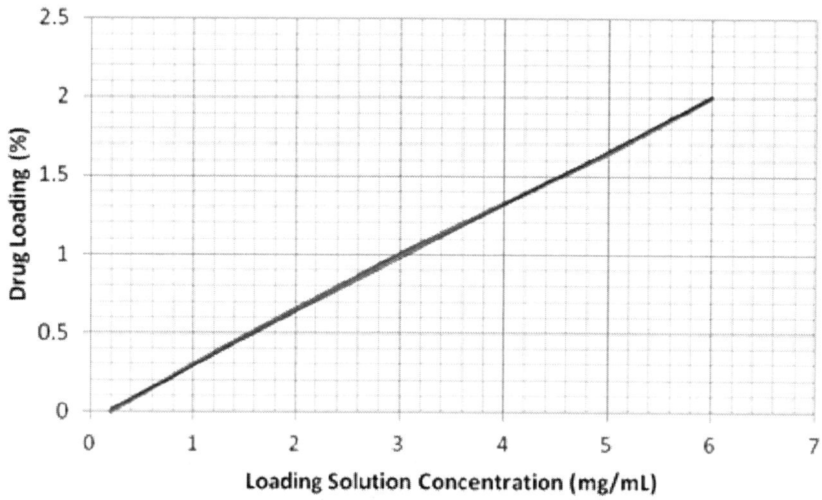


Using ATP levels as an indicator for tissue viability is not necessarily an accurate measure, however. Dowty et al., reported a 50% drop in the tissue ATP concentration during the initial 6 hours of the experiment without a corresponding drop in tissue permeability. Despite certain gradual changes, the buccal tissue seems to remain viable for a rather long period of time. Therefore, a decrease in ATP levels does not assure a drop in permeability characteristics of the tissue. The most meaningful method to assess tissue viability is the actual permeation experiment itself, if the drug permeability does not change during the time course of the study under the specific experimental conditions of pH and temperature, then the tissue is considered viable. Buccal cell cultures have also been suggested as useful *in vitro* models for buccal drug permeation and metabolism. However, to utilize these culture cells for buccal drug transport, the number of differentiated cell layers and the lipid composition of the barrier layers must be well characterized and controlled. This has not yet been achieved with the buccal cell cultures used thus far. Other than the low flux associated with buccal mucosal delivery, a major limitation of the buccal route of administration is the lack of dosage form retention at the site of absorption. Consequently, bioadhesive polymers have extensively been employed in buccal drug delivery systems. Bioadhesive polymers are defined as those polymers that can adhere onto a biological substrate. The term mucoadhesion is applied when the substrate is on the mucosal tissue. Polymers which can adhere to either hard or soft tissue have been used for many years in surgery and dentistry.

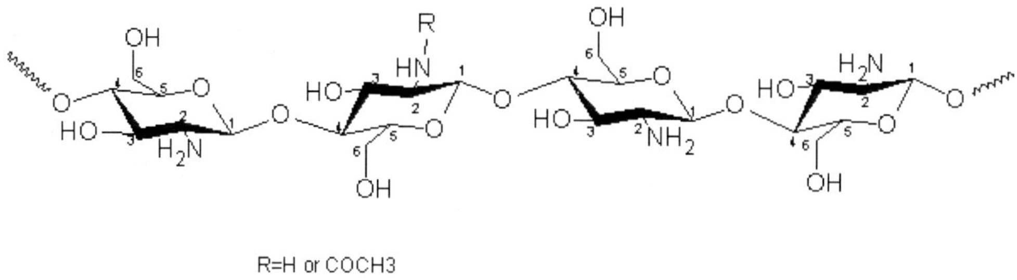


Chitosan

Diverse classes of polymers have been investigated for their potential use as mucoadhesives. These include synthetic polymers such as monomeric cyanoacrylate, polyacrylic acid, hydroxypropyl methylcellulose, and poly methacrylate derivatives. There are also naturally occurring polymers such as hyaluronic acid and chitosan.

Other synthetic polymers such as polyurethanes, epoxy resins, polystyrene, and natural-product cement have also been extensively investigated. A novel system developed and evaluated adhesive patches for buccal administration, consisting of two-ply laminates of an impermeable backing layer and a hydrocolloid polymer layer containing the drug. The polymers used HPC, HEC, PVP, and PVA. The integrity of the laminate was based on adhesive bonds between the hydrocolloid layer and an agarose layer grafted to one side of the backing layer sheet. Their work showed that among the cellulose ethers studied HEC and HPC possessed superior mucosal adhesion. Ishida et al., utilized similar materials in a lidocaine delivery system for a toothache. They used an HPC-CP powder mixture as a peripheral base aiming at strong adhesion, but an HPC-CP freeze-dried mixture was used as a core base. HPC and CP formed a complex during the freeze drying process, and this contributed to the ease of admixing by blockage of the functional group of CP. Guo used a two roll milling method to prepare a new bioadhesive polymer patch formulation for controlled drug delivery consisting of CP, PIB, and PIP. It was found that the surface properties of buccal patches were not only dependent on the CP content but also dependent on the PIB:PIP ratio. The strongest peel strength was found on buccal patches with a CP:PIB:PIP ratio of 50:43.75:6.25. In general, dosage forms designed for buccal administration should not cause irritation and should be small and flexible enough to be accepted by the patient. These requirements can be met by using hydrogels. Hydrogels are hydrophilic matrices that are capable of swelling when placed in aqueous media. Normally, hydrogels are cross-linked so that they would not dissolve in the medium and would only absorb water. When drugs are loaded into these hydrogels, as water is absorbed into the matrix, chain relaxation occurs and drug molecules are released through the spaces or channels within the hydrogel network. In a broader meaning of the term, hydrogels would also include water-soluble matrices that are capable of swelling in this aqueous media, as to include natural gums and cellulose derivatives. These 'pseudo-hydrogels' swell infinitely and the component molecules dissolve from the surface of the matrix. Drug release would then occur through the spaces or channels within the network as well as through the dissolution and/or the disintegration of the matrix. However applicability of hydroxypropyl cellulose (HPC) as a mucoadhesive agent, and determined that this high viscosity grade material to be a suitable adhesive for topical mucus membranes. They reported the combination of HPC and carbopol 934P (CP) to produce a preferable material for mucoadhesive dosage forms. They examined directly compressed tablets of these polymers by placing them on an agar gel bed. HPC tablets showed a slight adhesion but dissolved easily on the gel bed. On the other hand, CP tablets showed strong adhesion but the swollen CP tablets seemed too hard. The combination of HPC and CP provided the mucoadhesion and adequate softness to prepare the tablets. Satoh et al., had measured the bioadhesive property of tablets consisting of HPC and CP using mouse peritoneal membrane. The adhesive force of the HPC-CP tablet was affected by the mixing ratio of HPC and CP. The adhesion force showed a minimum value at the mixing ratio of 3:2 (HPC:CP) due to the formation of an inter-polymer complex between HPC and CP in the acidic pH range. Complex formation between CP and HPC seemed to suppress the interaction between molecules of hydrogel and the mucus membrane, and the adhesion force was therefore reduced the most at a mixture ratio of 1:4 (HPC/CP). Inter-polymer complexation and its effect on bioadhesive strength were also studied. They reported that CP shows strong complexation with poly(vinyl pyrrolidone) and hydroxypropyl cellulose, but very little with sodium carboxy methyl cellulose. The degree of complexation was higher at acidic pH and decreased with an increase in pH. Research reported on formulation and evaluation of buccoadhesive controlled release systems for the delivery of morphine sulfate. They prepared tablets by direct compression of carbomer and hydroxypropyl methyl cellulose (HPMC) and found the drug release behavior to be non-Fickian and also confirmed interpolymer complex formation between HPMC and carbomer in acidic pH medium. Also, some novel hydrogels were formed by the combination of natural gums, xantham gum, and locust bean gum, which are applicable in buccal delivery systems. Xantham gum is a natural gum obtained through fermentation of glucose by *Xanthamonas campestris*. Locust bean gum and xantham gum alone cannot form a hydrogel. However, when a mixture of these natural gums are dissolved in a neutral medium at 90°C and then cooled with ice for 30 min, a clear, strong hydrogel is formed. The mechanism of gel formation was reported to be the formation of a three-dimensional network from the interaction between the double helix structure of xantham gum and the straight molecular chain of locust bean gum. The gel strength of the hydrogels was affected by the mixing ratio of the gums, and the addition of sucrose improved the sustained release properties of the hydrogels. The hydrogel consisting of xantham gum and locust bean gum showed only a low mucoadhesion, but it can be applied to a buccal delivery system because of its safety, gel strength, sustained release properties and good comfortable feel within the mouth. A new system designed a bilayer patch (polytef-disk) consisting of protirelin for thyroid gland diagnosis. The patch had a backing layer of teflon and mucoadhesive layer of protirelin dispersed in hydroxyethylcellulose. The use of a unidirectional buccal patch which consisted of three layers: an impermeable backing layer, a rate limiting center membrane containing the drug, and a mucoadhesive layer containing bioadhesive polymer polycarbophil. The bioadhesive polymer swells, creating a flexible network through which diffusion of drug takes place. This patch was tested in dog buccal mucosa and was shown to remain in place for up to 17 hours without any obvious discomfort. The scientists examined a large number of polymers as to their bioadhesive potential and to derive meaningful information on the structural requirements for bioadhesion. They concluded that charged carboxylated polyanions are good potential bioadhesives for drug delivery. To understand the role of the carboxyl groups in mucoadhesion, acrylic acid-acrylamide random copolymers [P(AA-co-AM)] were synthesized and the adhesion strength of the cross-linked polymers to the gastric mucus layer as a function of pH, initial concentration of the cross-linking agent, degree of swelling, and carboxyl group density were examined. From the study on mucoadhesion of various P(AA-co-AM), it was found that at least 80% of the vinyl groups of the polymer must possess carboxyl groups in the protonated form. The dependence of mucoadhesion on pH and carboxyl-group density suggests that mucoadhesion occurs through hydrogen bonding. In addition, the density of the cross-linking agent significantly affected the mucoadhesion. As the density of the cross-linking agent is lowered, the mucoadhesive strength increases, although the density of carboxyl groups in the test surface area is reduced. It was concluded that for mucoadhesion to occur, polymers must have functional groups that are able to form hydrogen bonds above the critical concentration (80% for vinyl polymers), and the polymer chains should be flexible enough to form as many hydrogen bonds as possible. Similar results were achieved when an evaluation was done on the effects of poly(acrylic acid) (PAA) molecular weight, the concentration of cross-linking agent (sucrose) on swelling, and drug (metoclopramide) release characteristics of PAA (carbopol) hydrogels.

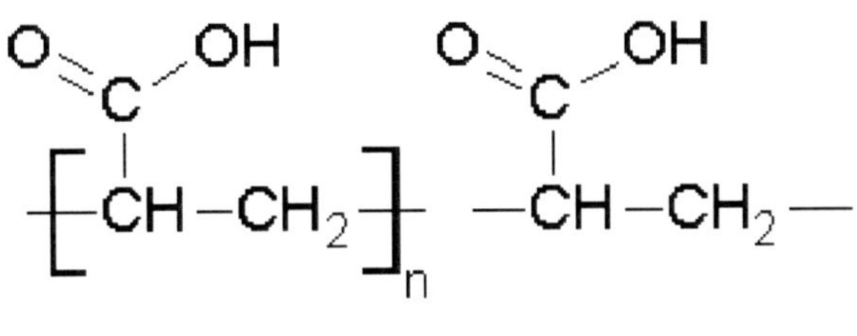


Poly(acrylic acid)

They reported that both of these factors in the interaction term had significant effects on hydrogel swelling and drug release. In particular, they found that increased sucrose concentration (high cross-linking density) led to reduced swelling and reduced drug release efficiency. The relationships between the structure and adhesion for the mucoadhesive polymers were investigated. Their study was based on an assumption that bioadhesion should possess two properties: (i) optimal polarity to make sure the polymer is "wetted" by the mucus, and (ii) optimal fluidity to allow for the mutual adsorption and interpenetration of polymer and mucus to take place. They studied acrylic polymer films, designed for buccal drug delivery. The films were made either through mixing or copolymerization of poly(butyl acrylate) and poly(acrylic acid).

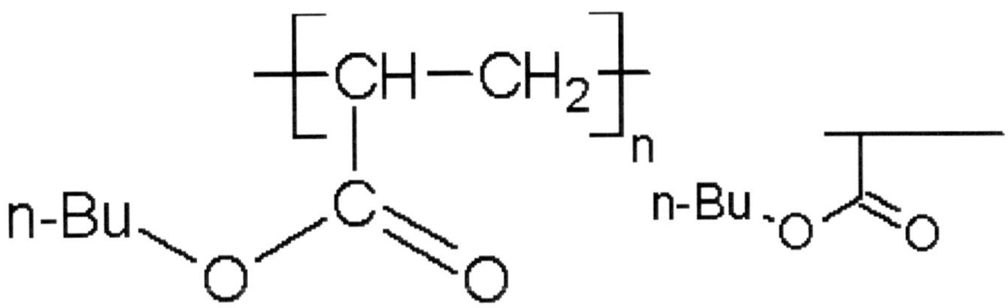


Poly(butyl acrylate)

In both cases, satisfactory mucoadhesion was found within a range of compositions optimized for surface polarity and the fluidity of the polymer film. A series of cross-linked, swellable polymers of acrylic acid and methacrylic acid were synthesized and evaluated, differing in charge densities and hydrophobicity. They found that an increase in the number of hydrophobic groups in the polymer structure reduced hydration whereas the density of the polymer was unaffected. Furthermore, polymers of acrylic acid loosely cross-linked (0.3 % w/w) with 3 different agents, divinyl glycol, 2,5-dimethyl-1,5-hexadiene, and divinyl benzene, showed the same degree of bioadhesion while poly(methacrylic acid) cross-linked with divinyl benzene showed a reduced bioadhesion. The small percent of cross-linking agent, irrespective of physicochemical properties, did not contribute substantially to bioadhesion, whereas the starting monomer had a large effect. The effect of pH on the bioadhesion of poly(acrylic acid) cross-linked with divinyl glycol was also studied at constant temperature, ionic strength, and osmolality. It was found that the polymer showed maximum adhesion at pH 5 and 6 and a minimum at pH 7. These studies show that the contribution of anionic polymer structural features to mucoadhesion when using 0.2% cross-linked copolymers of acrylic acid and methyl methacrylate. Their results showed that the expanded nature of both the interacting mucus and polymer networks influences the strength of mucoadhesion. They concluded that mucoadhesive polymer with the desired mucoadhesive strength can be designed by controlling the percentage of charged groups and the corresponding openness of the network. Absorption of drugs through the oral cavity was noted as early as 1847 by Sobrero, the discoverer of nitroglycerin, and systemic studies of oral cavity absorption were first reported. Since then, substantial effort has been focused on drug absorption from a drug delivery system in a particular region of the oral cavity. As a site for drug delivery, the oral cavity offers many advantages over other routes of drug administration. The mucosal lining of the oral cavity are readily accessible, robust, and heal rapidly after local stress or damage. Oral mucosal drug delivery systems can be localized easily and are well accepted by patients. Therefore, it is evident that the oral cavity can serve as a site for systemic drug delivery. Using copolymeric hydrogel discs of HEMA (monomer) and Polymeg® (macromer) a buccal mucoadhesive device for controlled release of buprenorphine was developed. The hydrogel containing a monomer:macromer ratio of 80:20 (w/w) yielded the best result both in terms of adhesion and drug release. The device was applied for a 3 hour application time and steady state levels were maintained for the time of application. Formulation of another buccal delivery system was reported from the 3M Company. The buccal patch device (Cydot; 3M Pharmaceutical, St. Paul, MN) consists of a flexible mucoadhesive matrix composed of a blend of poly(acrylic acid) (Carbopol 934P; B.F. Goodrich, Cleveland, OH) and poly(isobutylene) (Vistanex; Exxon Chemical Company, Houston, TX). The patch device is unidirectional with a polyurethane backing layer. The patch is intended for application to the upper gum and *in vivo* studies in human subjects and revealed effective bioadhesive characteristics for a 12 hour application. Several investigators have reported on the development of TmTs (transmucosal therapeutic systems) devices with a field-shaped bilayer design consisting of fast-release and sustained-release layers. The fast release layer contains PVP as the bioadhesive component and is designed to adhere to the buccal mucosa and the sustained release layer consists of a mixture of PVP and poly(acrylic acid) and is intended to adhere to the gingival mucosa. Most recently, a TmTs formulation was reported for the buccal delivery of LHRH (luteinizing hormone releasing hormone) with results indicating the feasibility of controlled release transmucosal delivery of the peptide drug. The total surface area of the oral cavity is about 100 $cm^2$. The mucosal membranes of the oral cavity can be divided into five regions: the floor of the mouth (sublingual), the buccal mucosa (cheeks), the gums (gingiva), the palatal mucosa, and the lining of the lips. In an attempt to enhance the intrinsic mucoadhesive properties of poly(acrylic acid), a series of novel copolymers of acrylic acid and poly ethyleneglycol monomethylether monomethacrylate [P(AA-co-PEG)] were designed and formulated. The addition of PEG into the polymer increased the potential for hydrogen bond formation, since the lone pair electrons of oxygen in the repeat unit ($CH_2CH_2O$) of PEG served as hydrogen bond acceptors. The surface properties of PAA for mucoadhesion were also improved by the PEG incorporation.

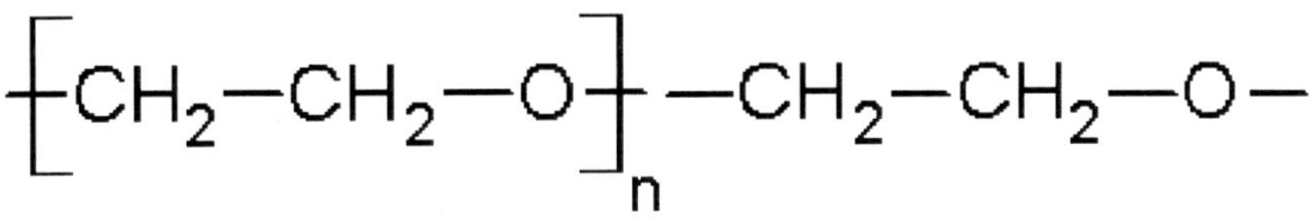

Poly(ethyleneglycol)

Using these copolymers, a patch device was prepared for buccal acyclovir delivery and the feasibility of such delivery was proven *in vitro* with the incorporation of sodium glycocholate as the permeation enhancer. These oral mucosal regions are different from each other in terms of anatomy, permeability to drug, and their ability to retain a system for a desired length of time. Although the buccal mucosa is less permeable than the sublingual mucosa and it does not yield a rapid onset of action as seen with sublingual delivery, mucosa of the buccal area has an expanse of smooth and relatively immobile surface, which is suitable for placement of a retentive system. These characteristics make the buccal mucosa a more appropriate site for prolonged systemic delivery of drugs. It has been shown that buccal route offers excellent opportunities for systemic delivery of drugs. In general, drug delivery through this route has the advantages of preventing the drug from degradation in the gastrointestinal tract, avoiding first-pass effect, and bypassing gastrointestinal absorption. The latter advantage is the rationale for designing a transbuccal drug delivery system of acyclovir to overcome the low oral bioavailability of the drug. Acyclovir is one of the most effective and selective agents against herpes viruses. It is widely used to treat herpes simplex virus (HSV) infections in both immunocompetent and immunocompromised patients. Current available dosage forms of acyclovir are intended for intravenous, oral, and topical administration. Systemic delivery of the drug following administration by these routes is far from being optimal, however. The oral absorption of acyclovir is dose dependent and highly variable with bioavailability ranging from 15 to 30%. The percutaneous penetration of acyclovir is poor. The limited solubility of acyclovir in water makes intramuscular administration rather difficult. Strongly alkaline (pH 10-11) solutions are required for intravenous infusion or bolus, which may result in thrombophlebitis or perivascular inflammation. There have been many attempts documented in literature for improving the physico-chemical properties of acyclovir by chemical modification. These include using highly water soluble esters of alkylamines and benzylcarbamates derivatives of acyclovir as prodrugs for ocular or parenteral formulations and aliphatic prodrugs of acyclovir to enhance its nasal transport, employing novel redox-based chemical targeting systems for improving intradermal delivery, as well as utilizing 6-deoxyacyclovir prodrug and 1-valyl ester prodrug, valacyclovir, for greater oral bioavailability. However, a logical approach to circumvent the bioavailability and physico-chemical problems of acyclovir is to deliver the drug through an alternative route. The purposes of this study were to design a buccal delivery system for acyclovir using a novel buccal adhesive and to evaluate the feasibility of transbuccal delivery of acyclovir through *in vitro* release and permeation studies using porcine buccal mucosa. The novel mucoadhesive hydrogels of copolymer [P(AA-co-PEG)] of acrylic acid (AA) and poly(ethylene glycol) monomethylether monomethacrylate (PEGMM) were used in designing the transbuccal delivery system.

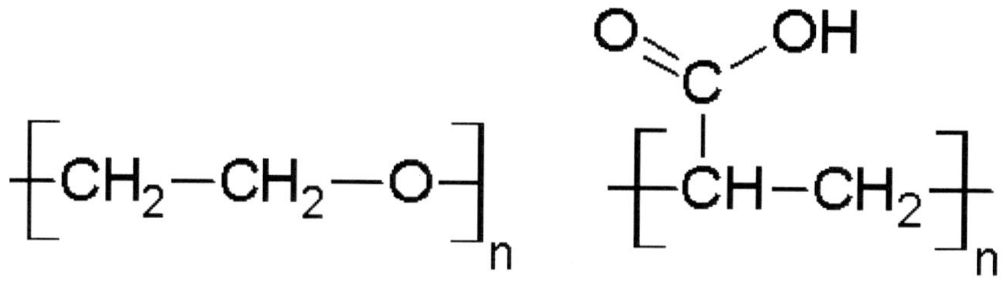


## REFERENCES


[1] Shojaei, A. H. (1998). Buccal mucosa as a route for systemic drug delivery: A review, *J. Pharm. Pharmaceut. Sci.,* (www.ualberta.ca/~csps) *1* (1),15-30.

[2] Miller, N. S. & Johnston, T. P. (2005). The use of mucoadhesive polymers in buccal drug delivery, *Adv. Drug Delivery Rev.*, *57*, 1666–1691.

[3] Lalla, J. K. & Gurnancy, R. A. (2002). Polymers for mucosal delivery-swelling and mucoadhesive evaluation, *Indian Drugs*, (5), 39.

[4] Mitra, A. K., Alur, H. H. & Johnston, T. P. (2002). *Peptides and Protein- Buccal Absorption, Encyclopedia of Pharmaceutical technology*, Marcel Dekker Inc., 2081-2093.

[5] Senel, S., Kremer, M., Nagy, K. & Squier, C. (2001). Delivery of bioactive peptides and proteins across oral (buccal) mucosa, *Curr. Pharm. Biotechnol.*, *2*, 175-186.

[6] Ghosh Tapash, K. & Pfister, W. R. Drug delivery to the oral cavity by Taylor and Francis, *Preclinical Assessment of Oral Mucosal Drug Delivery Systems*, p. 46.

[7] Bruschi, M. L. & de Freitas, O. (2005). Oral bioadhesive drug delivery systems, *Drug Dev. Ind. Pharm.*, *31*(3), 293-310.

[8] Wen, H. & Park, K. Oral controlled release formulation design and drug delivery, *Theory to Practice*, 169-183.

[9] Raj, B., Ravichandran, M., Li, X. & Jasti, B. R. (2007). Advances in buccal drug delivery, *Current Status In Buccal Drug Del. Syst.*, Vol. *5*, Issue 2.

[10] Vaughan, D. F. (2003). Pharmacokinetics of albuterol and butorphanol administered intravenously and via a Buccal Patch, Thesis Submitted, Graduate Studies of Texas A&M University. (*In Partial Fulfillment of the requirements for the Degree of Master of Science*, May 2003.)

[11] Harris, D. & Robinson, J. R. (1992). Drug delivery via the mucous membranes of the oral cavity, *J. Pharm. Sci.*, *81*, 1-10.

[12] Wise, D. L. *Handbook of Pharmaceutical Controlled Release Technology*, 255-265.

[13] Yajaman, S. & Bandyopadhyay, A. K. (2006). Buccal bioadhesive drug delivery- A promising option for orally less efficient drugs, *J. Controlled Rel.*, *114*, 15–40.

[14] Jain, N. K. *Controlled and Novel Drug Delivery,* 65-75, 371-377.

[15] Edgar, W. M. (1992). Saliva: Its secretion, composition and functions, *Brit. Dent. J.*, *172*, 305-312,

[16] *Indian J. Pharm. Sci.*, (2004). *66* (4), 371-536, 556-562.

[17] www.controlled drugdelivery.com.
[18] Savage, D. C. (1977). Microbial ecology of the gastrointestinal tract, *Ann. Rev. Microbiol.*, *31*, 107–133.
[19] Vyas, S. P. & Khar Roop, K. *Controlled Drug Delivery System*, 257-295.
[20] Salamat-Miller, N., Chittchang, M. & Johnston, T. P. (2005). The use of mucoadhesive polymers in buccal drug delivery, *Adv. Drug Delivery Rev.*, *57*(11), 1666-1691.
[21] Shivam, T. & Nishiprakash, J. (2011). Buccal control drug delivery system: A review, *IJPSR*, *2*(1), 27-38.
[22] Sevdsenel, A. (2001). A Hincal, drug permeation enhancement via buccal route: Possibilities and limitations. *J. Controlled Rel.*, *72*, 133-144.
[23] Hung Seng Ch, N. G., Park, H., Kelly, P. & Robinson, J. R. (1985). Bioadhesive polymers as platforms for oral controlled drug delivery-II: Synthesis and evaluation of some swelling water insoluble bioadhesive polymers. *J. Pharm. Sci.*, *74*(4), 399-409.

*Chapter 6*

# MUCOADHESIVE OCULAR DELIVERY SYSTEMS

The ARVO 2009 Summer Eye Research Conference (SERC 2009) on Ophthalmic Drug Delivery Systems was held July 31 and August 1, 2009, at the National Institutes of Health (NIH) in Bethesda, Maryland. The conference provided an opportunity to gather a diverse group of more than 200 experts from both academic ophthalmology and the ophthalmic pharmaceutical industry, including laboratory researchers and clinicians, to discuss recent advances in delivery systems that convey ocular drugs to the posterior segment and how these systems might be successfully used in commercial products. The two-day meeting comprised the following nine sessions: (1) eye anatomy and ocular barriers to drug transport, (2) the vitreous humor in drug delivery, (3) intravitreal drug delivery, (4) transscleral drug delivery for retinal diseases, (5) preclinical benchmarks for retinal drug therapy, (6) animal models for evaluating drug delivery systems, (7) topical therapy for retinal diseases, (8) clinical trials, and (9) new data from abstracts. This meeting was co-sponsored by the Association in Research in Vision and Ophthalmology (ARVO) and the NIH to expand and continue ongoing collaborative interchange and synergic endeavors that were addressed at an earlier conference, held May 4 and 5, 2007, and sponsored by the Pfizer Ophthalmics Research Institute Conference. SERC 2009 was organized by Cheryl L. Rowe-Rendleman, PhD (Omar Consulting Group, LLC), Michael R. Robinson, MD (Allergan Inc.), and Henry F. Edelhauser, PhD (Emory University). Edelhauser and Paul A. Sieving, MD, PhD, started off the meeting with introductory comments.

## CONVENTIONAL DELIVERY SYSTEMS

### Eye Drops

Drugs that are active in the eye or eye surface are widely administered in the form of solutions, emulsion, and suspension. Generally, eye drops are used only for anterior segment disorders since adequate drug concentrations are not reached in the posterior tissues using this drug delivery method. Various properties of eye drops like hydrogen ion concentration, osmolality, viscosity, and instilled volume can influence retention of a solution in the eye. Less than 5 % of the dose is typically absorbed after a topical drug administration is put into the eye. The dose is mostly absorbed into the systemic blood circulation via the conjunctival and nasal blood vessels. Ocular absorption is limited by the corneal epithelium, and it is only moderately increased by prolonged ocular contact. The reported maximal attainable ocular absorption is only about 10 % of the dose. When eye drops are administered in the inferior fornix of the conjunctiva, a very small amount of the dose reaches the intraocular tissues and a major fraction of the administered drug gets washed away with the lachrymal fluid or absorbed systemically in the nasolacrimal duct and pharyngeal sites.

### Ointment and Gels

Prolongation of drug contact time with the external ocular surface can be achieved using an ophthalmic ointment vehicle but, the major drawback of this sort of dosage is the blurring of vision and matting of eyelids can limit its use. Pilopine HS gel containing pilocarpine was used to provide sustain action over a period of 24 hours. A number of cases reported that these ointments and gels prolonged the corneal contact time for many of the drugs administered by topical ocular route. Thus, by this prolonged duration it increased the action and enhanced the ocular bioavailability of drugs.

### Ocuserts and Lacrisert

Ocular insert (Ocusert) is a sterile preparation that has a prolonged residence time, where the drug can provide a controlled release that is negligible or less affected by nasolacrimal damage. Inserts are available in different varieties depending upon their composition and applications. Lacrisert is a sterile rod-shaped device for the treatment of dry eye syndrome and keratitis sicca; and was introduced by Merck, Sharp, and Dohme in 1981. This treatment uses a solution of water on the cornea and conjunctiva to form a hydrophilic film that lubricates the cornea.

## Vesicular System

### Liposomes

Liposomes are biocompatible and biodegradable lipid vesicles made up of natural lipids and about 25–10 000 nm in diameter. The liposomes allow intimate contact with the corneal and conjunctival surfaces which is desirable for drugs that are poorly absorbed. Drugs that are poorly absorbed may have a low partition coefficient, or a poor solubility or those with medium to high molecular weights; and thus, increase the probability of ocular drug absorption. The corneal epithelium is thinly coated, with negatively charged mucin, to which the positive charged surface of the liposomes may bind.

### Niosomes and Discomes

The major limitations for liposomes are its chemical instability, oxidative degradation of phospholipids, cost, and purity of natural phospholipids. To avoid this, niosomes were developed since they are chemically stable as compared to liposomes and can entrap both hydrophobic and hydrophilic drugs. They are non-toxic and do not require special handling techniques. Niosomes are nonionic surfactant vesicles have potential applications in the delivery of hydrophobic or amphiphilic drugs. Vyas et al., reported that there was about 2.49 times increase in the ocular bioavailability of timolol maleate encapsulated in niosome as compared to timolol maleate solution. Non-ionic surface active agents based discoidal vesicles, known as (discomes), loaded with timolol maleate were formulated and characterized for their *in vivo* parameters. *In vivo* studies showed that discomes released the contents in a biphasic profile if the drug was loaded using a pH gradient technique. Discomes may act as a potential drug delivery carrier, as they possess the ability to release a drug in a sustained manner at the ocular site.

### Pharmacosomes

This term is used for pure drug vesicles formed by the amphiphilic drugs. Any drug possessing a free carboxyl group or an active hydrogen atom can be esterified (with or without a spacer group) to the hydroxyl group of a lipid molecule, thus generating an amphiphilic prodrug. The amphiphilic prodrug is converted to pharmacosomes on dilution with water. The pharmacosomes show greater shelf stability, facilitated transport across the cornea, and possesses a controlled release profile.

## Control Delivery Systems

### Implants

For chronic ocular diseases like cytomegalovirus (CMV) retinitis, implants are effective drug delivery system. Earlier nonbiodegradable polymers were used but they needed surgical procedures for insertion and removal. Presently biodegradable polymers such as Poly Lactic Acid (PLA) are safe and effective to deliver drugs in the vitreous cavity and show no toxic signs. Intravitreal implants of fluocinolone acetonide were developed for the treatment of posterior segment and reported to control the ocular inflammation in the retina.

### Iontophoresis

Iontophoresis directs current which drives ions into the cells or tissues. For iontophoresis, the ions of importance are the charged molecules of the drug. These positively charged molecules of drug are driven into the tissues at the anode and vice versa. Ocular iontophoresis delivery is not only fast, painless, and safe, but it can also deliver a high concentration of the drug to a specific site. Iontophoretic application regarding the use of antibiotics in eye not only increases their bactericidal activity but also reduce the severity of disease. Similarly, the application of anti-inflammatory agents can reduce any vision threatening side effects.

## Dendrimer

Dendrimers can be successfully used for different routes of drug administration and have better water-solubility, bioavailability, and biocompatibility. Vandamme et al., developed and evaluated poly (amidoamine) dendrimers containing fluorescein for controlled ocular drug delivery. They determined the influence of size, molecular weight, and number of amine, for carboxylate and hydroxyl surface groups in several series of dendrimers. The residence time was longer for the solutions containing dendrimers with carboxylic and hydroxyl surface groups.

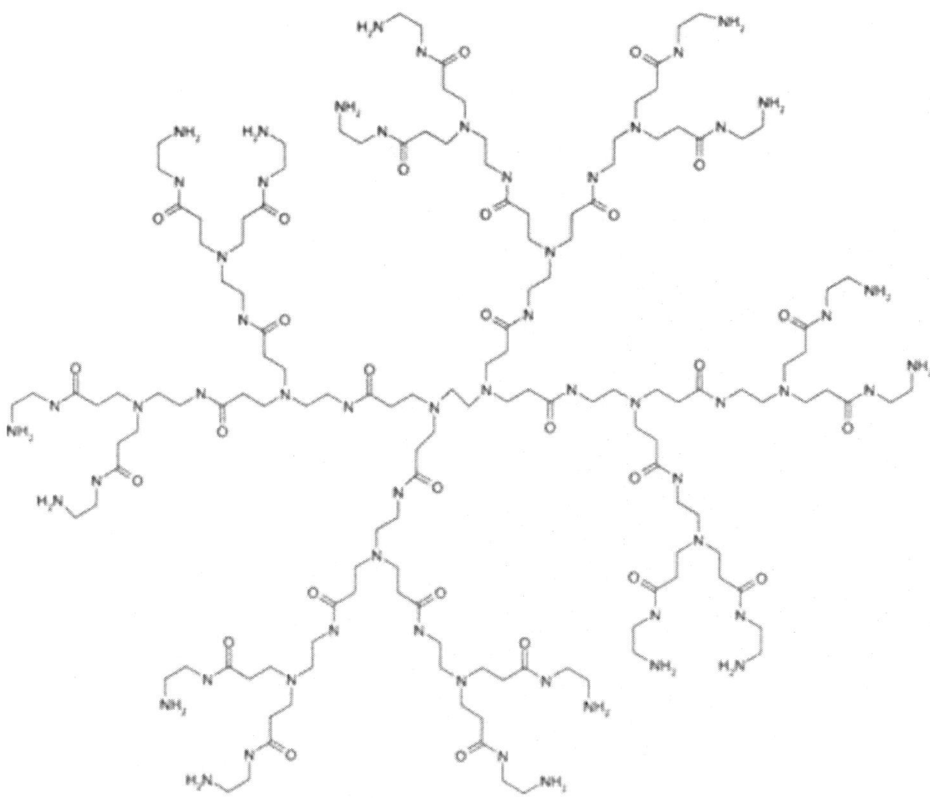

## Cyclodextrin

Cyclodextrins (CDs) are cyclic oligosaccharides capable of forming inclusion complexes with many guest molecules. CD complexes are reported to increase corneal permeation of drugs like dexamethasone, dexamethasone acetate, cyclosporine, and pilocarpine, which resulted in a higher bioavailability than the conventional eye drops. This complexation of CD does not interrupt the biological membrane compared to conventional permeation enhancer like benzalkonium chloride. Due to inclusion, the free drug is not available, so drugs with inherent irritant properties can be successfully delivered by this approach. CD molecules are inert in nature and were found to be nonirritant to the human and animal eye.

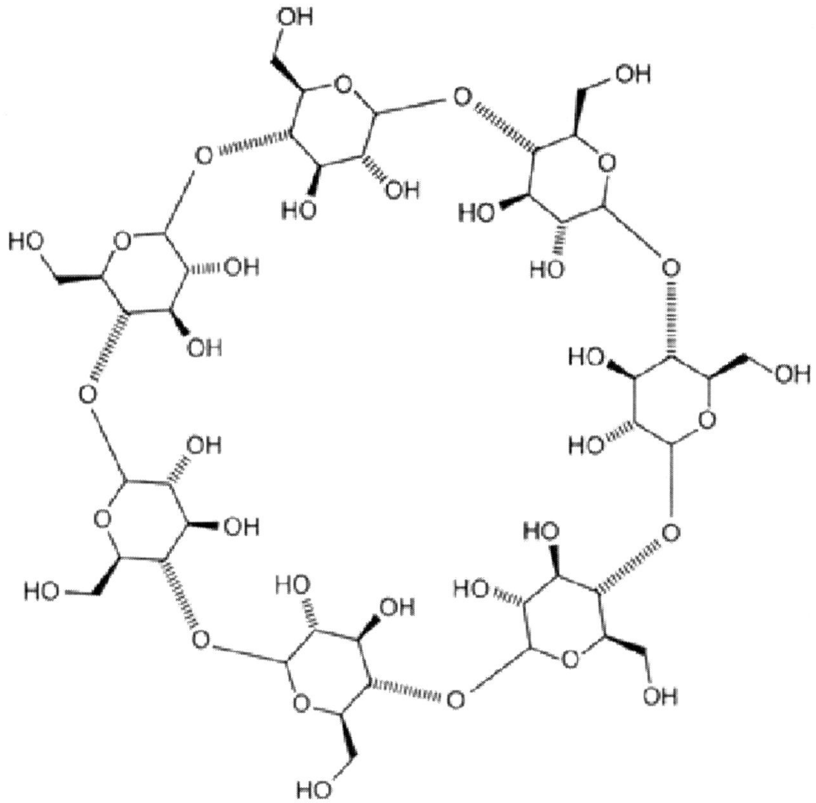


## Contact Lenses

Water soluble drugs soaked in drug solutions can be absorbed through contact lenses. The drug saturated contact lenses are placed in the eye which releases the drug in the eye for a long period of time.

For drug prolongation concerning the ocular residence time, hydrophilic contact lenses can be used. Greater penetration of fluorescein has been reported by Bionite lens made from hydrophilic polymer (2-hydroxy ethyl methacrylate) in human volunteers.

## Collagen Shield

The collagen shield basically consists of a cross-linked collagen, fabricated with fetal calf skin tissue and developed for the application as a corneal bandage to promote wound healing. Topically applied antibiotic is conjugated with the shield and used to promote the healing of corneal ulcers. Tear fluid makes these devices soft and form a thin pliable film which provides an dissolution rate of up to 10, 24, or 72 hours.

Because of its structural stability, good biocompatibility and biological inertness, collagen film proved as a potential carrier for ophthalmic drug delivery system. Collagen ophthalmic inserts are available for the delivery of pilocarpine to the eye.

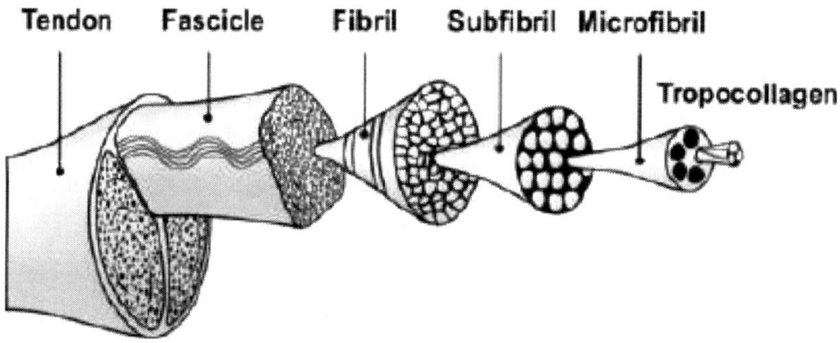


COLLAGEN STRUCTURE

## Microemulsion

Microemulsion is the dispersion of water and oil stabilized by using a surfactant and cosurfactant to reduce interfacial tension and is usually characterized by its small droplet size (100 nm), higher thermodynamic stability, and clear appearance. The most critical parameters to affect the stability of microemulsion would be in selecting the aqueous phase or the organic phase for the surfactant/cosurfactant systems.. The optimization of these components can result in a significant improvement in solubility of the drug molecule, e.g., indomethacin, chloramphenicol for eye diseases.

## Nanosuspensions

Nanosuspensions have emerged as a promising strategy for the efficient delivery of hydrophobic drugs. Simply because they can enhance not only the rate and extent of ophthalmic drug absorption but also the intensity of drug action with a significant extended duration in its drug effect. For commercial preparation of nanosuspensions, techniques like media milling and high pressure homogenization have been used. The higher drug level in the aqueous humor was reported using Eudragit™ RS100 nanosuspensions for the ophthalmic controlled delivery of ibuprofen.

## Microneedle

As an alternative to the topical route, researchers have developed a microneedle to deliver a drug to the posterior segment. The extent of lateral and transverse diffusion of sulforhodamine was reported to be similar across the sclera of a human cadaver. Microneedle had shown prominent *in vitro* penetration into the sclera and rapid dissolution of coating solution after insertion, while the *in vivo* drug level was found to be significantly higher than the level observed following topical drug administration like pilocarpine.

## Prodrugs

The ideal prodrugs for ocular therapy must not only be able to increase lipophilicity and have a high partition coefficient. But, the prodrug must also have high enzyme susceptibility. To the extent when after corneal penetration or within the cornea, it can either be chemically or enzymatically metabolized to the active parent compound. The partition coefficient of ganciclovir was found to increase when using an acyl ester prodrug. This substantially increased the amount of drug penetration to the cornea due to increased susceptibility of the ganciclovir esters to undergo hydrolysis by esterases in the cornea.

## Penetration Enhancers

The transport of a drug across the cornea is increased by increasing the permeability through corneal epithelial membranes. For this purpose penetration enhancers can be used. Examples of these penetration enhancers include actin filament inhibitors, surfactants, bile salts, chelators, and organic compounds. The selection process of an enhancer is critical due to the unique characteristics and great sensitivity of the corneal conjunctival tissues. Penetration enhancers themselves can penetrate the eye and may lead to unknown toxicological complications e.g., benzalkonium chloride (BAC) was found to accumulate in the cornea for days.

## Mucoadhesive Polymers

They are basically macromolecular hydrocolloids with plentiful hydrophilic functional groups, such as hydroxyl, carboxyl, amide and sulphate with the capability for establishing electrostatic interactions. A mucoadhesive drug formulation for the treatment of glaucoma was developed using a highly potent beta blocker drug, levobetaxolol (LB) hydrochloride and partially neutralized poly acrylic acid (PAA). Complexes were prepared with varying degrees of drug loading, such that the same PAA chain would have free -COOH groups for mucoadhesion along with ionic complexes of LB with COO- groups. Thin films of the complexes dissociated to release the drug by ion exchange within the synthetic tear fluid.

## Phase Transition Systems/*In situ* Gel System

Phase transition of the formulation from the liquid form to the gel or solid phase occurs when these systems are instilled into the cul-de-sac of eye which led to an increase the viscosity of the drug formulation in the precorneal region. The results showed an increased bioavailability, due to slower drainage from the cornea. These systems can be influenced by pH, temperature, or by ion activation. A sol to gel system with mucoadhesive property to deliver the steroid fluorometholone to the eye was prepared by Middleton and Robinson.

### Particulates (Nanoparticles and Microparticles)

The maximum size limit for microparticles when using ophthalmic administration is about 5-10 mm. When the microparticle size is beyond 10mm, the result causes a scratching feeling in the eye upon ocular instillation. That is why microspheres and nanoparticles are promising drug carriers for ophthalmic application. Nanoparticles are prepared using bioadhesive polymers to provide sustained effect to the entrapped drugs. An optimal corneal penetration of the encapsulated drug was reported in the presence of bioadhesive polymer chitosan. Similarly, poly butyl cyanoacrylate nanoparticles, containing pilocarpine into collagen shields, showed greater retention and activity characteristics with respect to the controls. Nanospheres made up of poly lactic acid (PLA) coated with poly ethylene glycol (PEG) shown better efficacy compared to conventional dosage form of acyclovir for the treatment of ocular viral infections. Microspheres of poly lacto gylcolic acid (PLGA) for topical ocular delivery of a peptide drug vancomycin were prepared by an emulsification/ spray-drying technique.

## Advanced Delivery Systems

### Cell Encapsulation

The entrapment of immunologically isolated cells with hollow fibers or microcapsules before their administration into the eye is called Encapsulated Cell Technology (ECT) which enables the controlled, continuous, and long-term delivery of therapeutic proteins directly to the posterior regions of the eye. The polymer implant containing genetically modified human RPE cells secretes ciliary neurotrophic factor into the vitreous humor of the patients' eyes. ECT can potentially serve as a delivery system for chronic ophthalmic diseases like neuroprotection in glaucoma, anti-angiogenesis in choroidal neovascularization, and anti-inflammatory factors for uveitis.

### Gene Therapy

Along with tissue engineering, gene therapy approaches stand on the front line of advanced biomedical research to treat blindness arising from corneal diseases, which are second only to cataract as the leading cause of vision loss. Several kinds of viruses including adenovirus, retrovirus, adeno-associated virus, and herpes simplex virus, have been manipulated for use in gene transfer and gene therapy applications. Topical delivery to the eye is the most expedient way of ocular gene delivery. However, the challenge of obtaining substantial gene expression following topical administration has led to the prevalence of invasive ocular administration. Retroviral vectors have been widely used due to their high efficacy; however, they do not have the ability to transduce nondividing cells, which restricts their clinical use. The advanced delivery systems that prolong the contact time with the surface of the eye may enhance transgene expression, thereby facilitate non-invasive administration.

## Stem Cell Therapy

Emerging cell therapies for the restoration of sight have focused on two areas of the eye that are critical for visual function, the cornea and the retina. Current strategy for management of ocular conditions consists of eliminating the injurious agent or attempting to minimize its effects. The most successful ocular application has been the use of limbal stem cells, transplanted from a source other than the patient for the renewal of corneal epithelium. The sources of limbal cells include donors, autografts, cadaver eyes, and (recently) cells grown in culture. Stem-cell therapy has demonstrated great success for certain maladies of the anterior segment.

## Protein and Peptide Therapy

Delivery of therapeutic proteins/peptides has received a great attention over the last few years. The intravitreous injection of ranibizumab is one such example. The design of optimized methods for the sustained delivery of proteins and to predict the clinical effects of new compounds when administered in the eye; does require the basic knowledge of proteins and peptides. However, several limitations such as membrane permeability, large size, metabolism, and solubility can restrict their efficient delivery. A number of approaches have been used to overcome these limitations. Poor membrane permeability of hydrophilic peptides may be improved by structurally modifying the compound, thus increasing their membrane permeability. Ocular route is not the preferred route for systemic delivery of such large molecules. Immunoglobulin G has been effectively delivered to the retina by a trans-scleral route with insignificant systemic absorption.

## Scleral Plug Therapy

A scleral plug can be implanted using a simple procedure at the pars plana region of eye. The sclera plug is made of biodegradable polymers and drugs, and it gradually releases effective drug doses for several months upon biodegradation. The release profiles vary with the kind of polymers used, their molecular weights, and the amount of drug in the plug. The plugs are effective for treating vitreoretinal diseases such as proliferative vitreoretinopathy, cytomegalovirus retinitis responds to repeated intravitreal injections, and for vitreoretinal disorders that require vitrectomy.

## siRNA Therapy

For various angiogenesis-related diseases, the use of siRNA is considered as a promising approach. Feasibility of using siRNA for treatment of choroidal neovascularization has been demonstrated using siRNA directed against vascular endothelial growth factor (VEGF) or VEGF receptor 1 (VEGFR1), and both of these approaches are being tested in clinical trials. Topical delivery of siRNAs directed against VEGF or its receptors has also been shown to suppress corneal neovascularization. A siRNA has become a valuable tool to explore the potential role of various genes in ocular disease processes. It appears that siRNAs may be valuable in the pathogenesis and development of new treatments for several ocular diseases, based on *in vivo* and *in vitro* studies. However, its use *in vivo* remains problematic, largely due to unresolved difficulties in targeting delivery of the siRNA to the tumor cells. Viral gene delivery is very efficient; however, it currently lacks adequate selectivity for the target cell type. New encapsulated siRNA have been developed using liposomes, coupled-antibodies or others polymer vesicles. Therapeutic approach to use siRNA provides a major new class of drugs that will shed new light on modern medicine.

## Oligonucliotide Therapy

Oligonucleotide (ON) therapy is based on the principle of blocking the synthesis of cellular proteins by interfering with either the transcription of DNA to mRNA or the translation of mRNA to proteins. Among several mechanisms by which antisense molecules disrupt gene expression and inhibit protein synthesis, the ribonuclease H mechanisms are the most important. A number of factors have been determined to contribute to the efficacy of antisense ON. One primary consideration is the length of the ON species. Lengths of 17– 25 bases have been shown to be optimal, as longer ONs have the potential to partially hybridize with nontarget RNA species. Biological stability is the major barrier to consider when delivering both DNA and RNA oligonucleotides to cells. Protection from nuclease action has been achieved by modification of phosphate backbones, sugar moiety, and bases.

## Aptamer

Aptamers are oligonucleotide ligands that are used for high-affinity binding to molecular targets. They are isolated from complex libraries of synthetic nucleic acid by an iterative process of adsorption, recovery, and reamplification. They bind with the target molecules at a very low level with high specificity. One of the earliest aptamers studied structurally was the 15 mer DNA aptamer against thrombin, d(GGTTGGTGTGGTTGG). Pegaptanib sodium (Macugen; Eyetech Pharmaceuticals/Pfizer) is an RNA aptamer directed against VEGFb165, where VEGF isoform primarily responsible for pathological ocular neovascularization and vascular permeability.

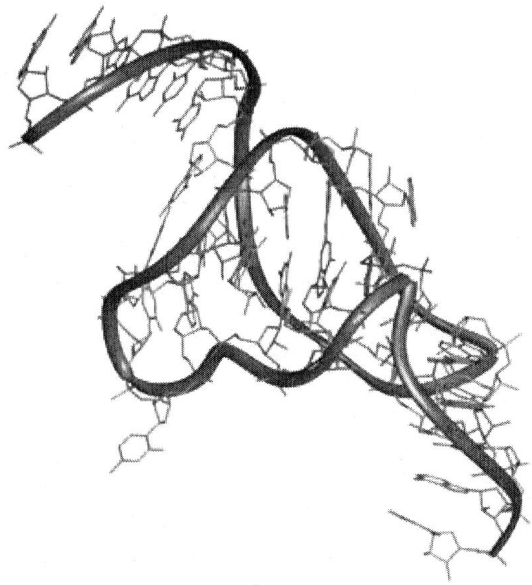

# RNA APTAMER

## Ribozyme Therapy

RNA enzymes or ribozymes are a relatively new class of single-stranded RNA molecules capable of assuming three-dimensional formations and exhibiting catalytic activity that induces site-specific cleavage, ligation, and polymerization of nucleotides involving RNA or DNA. They function by binding to the target RNA moiety through Watson-Crick base pairing and can inactivate it by cleaving the phosphodiester backbone at a specific cutting site. A disease named, Autosomal dominated retinitis pigmentosa (ADRP) is caused by mutations in genes that produce mutated proteins, leading to the apoptotic death of photoreceptor cells. Lewin and Hauswirth have worked on the delivery of ribozymes in ADRP in rats that shows promise for ribozyme therapy in many other autosomal dominant eye diseases, including glaucoma.

# REFERENCES

[1] Ananthula, H. K., Vaishya, R. D., Barot, M. & Mitra, A. K. (2009). Duane's ophthalmology. In: W. Tasman and E.A. Jaeger, (eds.), *Bioavailability*, Philadelphia: Lippincott Williams and Wilkins.

[2] Gipson, I. K. & Argueso, P. (2003). Role of mucin in the function of the corneal and conjunctival epithelia. *Int. Rev. Cytol.*, *231*, 1–49.

[3] Ahmed, I. (2003). The noncorneal route in ocular drug delivery. In: A.K. Mitra, (ed.), *Ophthalmic Drug Delivery Systems*, New York: Marcel Dekker, 335–363.

[4] Klyce, S. D. & Crosson, C. E. (1985). Transport processes across the rabbit corneal epithelium: A review. *Curr. Eye Res.*, *4*(4), 323–331.

[5] McLaughlin, B. J., Caldwell, R. B., Sasaki, Y. & Wood, T. O. (1985). Freeze fracture quantitative comparison of rabbit corneal epithelial and endothelial membranes. *Curr. Eye Res.*, *4*(9), 951–961.

[6] Barar, J., Javadzadeh, A. R. & Omidi, Y. (2008). Ocular novel drug delivery: Impacts of membranes and barriers. *Expert Opin. Drug Deliv.*, *5*(5), 567–581.

[7] Sunkara, G. K. U. (2003). Membrane transport processes in the eye. In: A.K. Mitra, (ed.), *Ophthalmic Drug Delivery Systems*, New York: Marcel Dekker, 13–58.

[8] Saha, P., Kim, K. J. & Lee, V. H. (1996). A primary culture model of rabbit conjunctival epithelial cells exhibiting tight barrier properties. *Curr. Eye Res.*, *15*(12), 1163–1169.

[9] Geroski, D. H. & Edelhauser, H. F. (2001). Trans-scleral drug delivery for posterior segment disease. *Adv. Drug Deliv. Rev.*, *52*(1), 37–48.

[10] Kim, S. H., Lutz, R. J., Wang, N. S. & Robinson, M. R. (2007). Transport barriers in trans-scleral drug delivery for retinal diseases. *Ophthalmic Res.*, *39*(5), 244–254.

[11] Pitkanen, L., Ranta, V. P., Moilanen, H. & Urtti, A. (2005). Permeability of retinal pigment epithelium: Effects of permeant molecular weight and lipophilicity. *Invest. Ophthalmol. Vis. Sci.*, *46*(2), 641–646.

[12] Urtti, A. (2006). Challenges and obstacles of ocular pharmacokinetics and drug delivery. *Adv. Drug Deliv. Rev.*, *58*(11), 1131–1135.

[13] Kim, J. H., Kim, K. W., Kim, M. H. & Yu, Y. S. (2009). Intravenously administered gold nanoparticles pass through the blood-retinal barrier depending on the particle size & induce no retinal toxicity. *Nanotechnology*, *20(*50), 505101.

[14] Zhu, C., Zhang, Y. & Pardridge, W. M. (2002). Widespread expression of an exogenous gene in the eye after intravenous administration. *Invest. Ophthalmol. Vis. Sci.*, *43*(9), 3075–3080.

[15] Singh, S. R., Grossniklaus, H. E., Kang, S. J., Edelhauser, H. F., Ambati, B. K. & Kompella, U. B. (2009). Intravenous transferrin, RGD peptide and dual-targeted nanoparticles enhance anti-VEGF intraceptor gene delivery to laser-induced CNV. *Gene Ther.*, *16*(5), 645–659.

[16] Suzuki, T., Uno, T., Chen, G. & Ohashi, Y. (2008). Ocular distribution of intravenously administered micafungin in rabbits. *J. Infect. Chemother.*, *14*(3), 204–207.

[17] Regnier, A., Schneider, M., Concordet, D. & Toutain, P. L. (2008). Intraocular pharmacokinetics of intravenously administered marbofloxacin in rabbits with experimentally induced acute endophthalmitis. *Am. J. Vet. Res.*, *69*(3), 410–415.

[18] Goldblum, D., Rohrer, K., Frueh, B. E., Theurillat, R., Thormann, W. & Zimmerli, S. (2002). Ocular distribution of intravenously administered lipid formulations of amphotericin B in a rabbit model. *Antimicrob. Agents Chemother.*, *46*(12), 3719–3723.

[19] Santulli, R. J., Kinney, W. A., Ghosh, S., et al., Studies with an orally bioavailable alpha V integrin antagonist in animal models of ocular vasculopathy: Retinal neovascularization in mice and retinal vascular permeability in diabetic rats. *J. Pharmacol. Exp. Ther.*, *324*(3), 894–901.

[20] Shirasaki, Y., Miyashita, H. & Yamaguchi, M. (2006). Exploration of orally available calpain inhibitors. Part 3: Dipeptidyl alpha-ketoamide derivatives containing pyridine moiety. *Bioorg. Med. Chem.*, *14*(16), 5691–5698.

[21] Kampougeris, G., Antoniadou, A., Kavouklis, E., Chryssouli, Z. & Giamarellou, H. (2005). Penetration of moxifloxacin into the human aqueous humor after oral administration. *Brit. J. Ophthalmol.*, *89*(5), 628–631.
[22] Sakamoto, H., Sakamoto, M., Hata, Y., Kubota, T. & Ishibashi, T. (2007). Aqueous and vitreous penetration of levofloxacin after topical and/or oral administration. *Eur. J. Ophthalmol.*, *17*(3), 372–376.
[23] Shirasaki, Y. (2008). Molecular design for enhancement of ocular penetration. *J. Pharm. Sci.*, *97*(7), 2462–2496.
[24] Kaur, I. P., Smitha, R., Aggarwal, D. & Kapil, M. (2002). Acetazolamide: Future perspective in topical glaucoma therapeutics. *Int. J. Pharm.*, *248*(1–2), 1–14.
[25] Coppens, M., Versichelen, L. & Mortier, E. (2002). Treatment of postoperative pain after ophthalmic surgery. *Bull Soc. Belge Ophtalmol.*, (285), 27–32.
[26] Samtani, S., Amaral, J., Campos, M. M., Fariss, R. N. & Becerra, S. P. (2009). Doxycycline-mediated inhibition of choroidal neovascularization. *Invest. Ophthalmol. Vis. Sci.*, *50*(11), 5098–5106.
[27] Rajpal, S. A., Azad, R. V., et al. (2009). Evaluation of vitreous levels of gatifloxacin after systemic administration in inflamed and non-inflamed eyes. *Acta Ophthalmol.*, *87*(6), 648–652.
[28] Smith, V. A., Khan-Lim, D. Anderson, L., Cook, S. D. & Dick, A. D. (2008). Does orally administered doxycycline reach the tear film? *Brit. J. Ophthalmol.*, *92*(6), 856–859.
[29] Chong, D. Y., Johnson, M. W., Huynh, T. H., Hall, E. F., Comer, G. M. & Fish, D. N. (2009). Vitreous penetration of orally administered famciclovir. *Am. J. Ophthalmol.*, *148*(1), 38–42.
[30] Takahashi, K., Saishin, Y., King, A. G., Levin, R. & Campochiaro, P. A. (2009). Suppression and regression of choroidal neovascularization by the multi-targeted kinase inhibitor pazopanib. *Arch. Ophthalmol.*, *127*(4), 494–499.
[31] Kokke, K. H., Morris, J. A. & Lawrenson, J. G. (2008). Oral omega-6 essential fatty acid treatment in contact lens associated dry eye. *Cont. Lens Anterior. Eye*, *31*(3), 141–146.
[32] Ghate, D. & Edelhauser, H. F. (2006). Ocular drug delivery. *Expert Opin Drug Deliv.*, *3*(2), 275–287.
[33] Hosseini, K., Matsushima, D., Johnson, J., et al. (2008). Pharmacokinetic study of dexamethasone disodium phosphate using intravitreal, subconjunctival, and intravenous delivery routes in rabbits. *J. Ocul. Pharmacol. Ther.*, *24*(3), 301–308.
[34] Weijtens, O., Feron, E. J., Schoemaker, R. C., et al. (1999). High concentration of dexamethasone in aqueous and vitreous after subconjunctival injection. *Am. J. Ophthalmol.*, *128*(2), 192–197.
[35] Kim, S. H., Csaky, K. G., Wang, N. S. & Lutz, R. J. (2008). Drug elimination kinetics following subconjunctival injection using dynamic contrast-enhanced magnetic resonance imaging. *Pharm. Res.*, *25*(3), 512–520.
[36] Prausnitz, M. R. & Noonan, J. S. (1998). Permeability of cornea, sclera, and conjunctiva: A literature analysis for drug delivery to the eye. *J. Pharm. Sci.*, *87*(12), 1479–1488.
[37] Mitra, A. K., Anand, B. S. & Duvvuri, S. (2006). Drug delivery to the eye. In: J. Fischbarg, (ed.), *The Biology of Eye*, New York: Academic Press, 307–351.

[38] Pitkanen, L., Ruponen, M., Nieminen, J. & Urtti, A. (2003). Vitreous is a barrier in nonviral gene transfer by cationic lipids and polymers. *Pharm. Res.*, *20*(4), 576–583.

[39] Peeters, L., Sanders, N. N,, Braeckmans, K., et al. (2005). Vitreous: A barrier to nonviral ocular gene therapy. *Invest. Ophthalmol. Vis. Sci.*, *46*(10), 3553–3561.

[40] Kim, H., Robinson, S. B. & Csaky, K. G. (2009). Investigating the movement of intravitreal human serum albumin nanoparticles in the vitreous and retina. *Pharm. Res.*, *26*(2), 329–337.

[41] Dalkara, D., Kolstad, K. D., Caporale, N., et al. (2009). Inner limiting membrane barriers to AAV-mediated retinal transduction from the vitreous. *Mol. Ther.*, *17*(12), 2096–2102.

[42] Schoenwald, R. D., Tandon, V., Wurster, D. E. & Barfknecht, C. F. (1998). Significance of melanin binding and metabolism in the activity of 5-acetoxyacetylimino-4-methyl-delta2–1, 3, 4,-thiadiazolin e-2-sulfonamide. *Eur. J. Pharm. Biopharm.*, *46*(1), 39–50.

[43] Larsson, B. S. (1993). Interaction between chemicals and melanin. *Pigment Cell Res.*, *6*(3), 127–133.

[44] Leblanc, B., Jezequel, S., Davies, T., Hanton, G. & Taradach, C. (1998). Binding of drugs to eye melanin is not predictive of ocular toxicity. *Regul. Toxicol. Pharmacol.*, *28*(2), 124–132.

[45] Pitkanen, L., Ranta, V. P., Moilanen, H. & Urtti, A. (2007). Binding of betaxolol, metoprolol and oligonucleotides to synthetic and bovine ocular melanin, and prediction of drug binding to melanin in human choroid-retinal pigment epithelium. *Pharm. Res.*, *24*(11), 2063–2070.

[46] Salminen, L., Imre, G. & Huupponen, R. (1985). The effect of ocular pigmentation on intraocular pressure response to timolol. *Acta Ophthalmol. Suppl.*, *173*, 15–18.

[47] Cheruvu, N. P. & Kompella, U. B. (2006). Bovine and porcine transscleral solute transport: Influence of lipophilicity and the Choroid- Bruch's layer. *Invest. Ophthalmol. Vis. Sci.*, *47*(10), 4513–4522.

[48] Gaudana, R., Jwala, J., Boddu, S. H. & Mitra, A. K. (2009). Recent perspectives in ocular drug delivery. *Pharm. Res.*, *26*(5), 1197–1216.

[49] Mannermaa, E., Vellonen, K. S. & Urtti, A. (2006). Drug transport in corneal epithelium and blood-retina barrier: Emerging role of transporters in ocular pharmacokinetics. *Adv. Drug Deliv. Rev.*, *58*(11), 1136–1163.

[50] Dey, S., Anand, B. S., Patel, J. & Mitra, A. K. (2003). Transporters/receptors in the anterior chamber: Pathways to explore ocular drug delivery strategies. *Expert Opin. Biol. Ther.*, *3*(1), 23–44.

[51] Kawazu, K., Yamada, K., Nakamura, M. & Ota, A. (1999). Characterization of cyclosporin-A transport in cultured rabbit corneal epithelial cells: P-glycoprotein transport activity and binding to cyclophilin. *Invest. Ophthalmol. Vis. Sci.*, *40*(8), 1738–1744.

[52] Dey, S., Patel, J., Anand, B. S., et al., (2003). Molecular evidence and functional expression of P-glycoprotein (MDR1) in human and rabbit cornea and corneal epithelial cell lines. *Invest. Ophthalmol. Vis. Sci.*, *44*(7), 2909–2918.

[53] Dey, S., Gunda, S. & Mitra, A. K. (2004). Pharmacokinetics of erythromycin in rabbit corneas after single-dose infusion: Role of P-glycoprotein as a barrier to *in vivo* ocular drug absorption. *J. Pharmacol. Exp. Ther.*, *311*(1), 246–255.

[54] Saha, P., Yang, J. J. & Lee, V. H. (1998). Existence of a p-glycoprotein drug efflux pump in cultured rabbit conjunctival epithelial cells. *Invest. Ophthalmol. Vis. Sci.*, *39*(7),1221–1226.

[55] Yang, J. J., Kim, K. J. & Lee, V. H. (2000). Role of P-glycoprotein in restricting propranolol transport in cultured rabbit conjunctival epithelial cell layers. *Pharm. Res.*, *17*(5), 533–538.

[56] Kennedy, B. G. & Mangini, N. J. (2002). P-glycoprotein expression in human retinal pigment epithelium. *Mol. Vis.*, *8*, 422–430.

[57] Duvvuri, S., Gandhi, M. D. & Mitra, A. K. (2003). Effect of P-glycoprotein on the ocular disposition of a model substrate, quinidine. *Curr. Eye Res.*, *27*(6), 345–353.

[58] Constable, P. A., Lawrenson, J. G., Dolman, D. E., Arden, G. B. & Abbott, N. J. (2006). P-Glycoprotein expression in human retinal pigment epithelium cell lines. *Exp. Eye Res.*, *83*(1), 24–30.

[59] Vellonen, K. S., Mannermaa, E., Turner, H., et al. (2010). Effluxing ABC transporters in human corneal epithelium. *J. Pharm. Sci.*, *99*(2), 1087–1098.

[60] Zhang, T., Xiang, C. D., Gale, D., Carreiro, S., Wu, E. Y. & Zhang, E. Y. (2008). Drug transporter and cytochrome P450 mRNA expression in human ocular barriers: Implications for ocular drug disposition. *Drug Metab. Dispos.*, *36*(7), 1300–1307.

[61] Karla, P. K., Quinn, T. L., Herndon, B. L., Thomas, P., Pal, D. & Mitra, A. (2009). Expression of multidrug resistance associated protein 5 (MRP5) on cornea and its role in drug efflux. *J. Ocul. Pharmacol. Ther.*, *25*(2), 121–132.

[62] Karla, P. K., Pal, D., Quinn, T. & Mitra, A. K. (2007). Molecular evidence and functional expression of a novel drug efflux pump (ABCC2) in human corneal epithelium and rabbit cornea and its role in ocular drug efflux. *Int. J. Pharm.*, *336*(1), 12–21.

[63] Yang, J. J., Ann, D. K., Kannan, R. & Lee, V. H. (2007). Multidrug resistance protein 1 (MRP1) in rabbit conjunctival epithelial cells: Its effect on drug efflux and its regulation by adenoviral infection. *Pharm. Res.*, *24*(8), 1490–1500.

[64] Aukunuru, J. V., Sunkara, G., Bandi, N., Thoreson, W. B. & Kompella, U. B. (2001). Expression of multidrug resistance-associated protein (MRP) in human retinal pigment epithelial cells and its interaction with BAPSG, a novel aldose reductase inhibitor. *Pharm. Res.*, *18*(5), 565–572.

[65] Karla, P. K., Earla, R., Boddu, S. H., Johnston, T. P., Pal, D. & Mitra, A. (2009). Molecular expression and functional evidence of a drug efflux pump (BCRP) in human corneal epithelial cells. *Curr. Eye Res.*, *34*(1), 1–9.

[66] Becker, U. Ehrhardt, C., Daum, N., et al. (2007). Expression of ABC transporters in human corneal tissue and the transformed cell line, HCE-T. *J. Ocul. Pharmacol. Ther.*, *23*(2), 172–181.

[67] Katragadda, S., Talluri, R. S., Pal, D. & Mitra, A. K. (2005). Identification and characterization of a Na+-dependent neutral amino acid transporter, ASCT1, in rabbit corneal epithelial cell culture and rabbit cornea. *Curr. Eye Res.*, *30*(11), 989–1002.

[68] Dun, Y., Mysona, B., Itagaki, S., Martin-Studdard, A., Ganapathy, V. & Smith, S. B. (2007). Functional and molecular analysis of D-serine transport in retinal Muller cells. *Exp. Eye Res.*, *84*(1), 191–199.

[69] Jain-Vakkalagadda, B., Pal, D., Gunda, S., Nashed, Y., Ganapathy, V. & Mitra, A. K. (2004). Identification of a Na+-dependent cationic and neutral amino acid transporter, B(0,+), in human and rabbit cornea. *Mol. Pharm.*, *1*(5), 338–346.

[70] Hosoya, K., Horibe, Y., Kim, K. J. & Lee, V. H. (1997). Na(+)-dependent Larginine transport in the pigmented rabbit conjunctiva. *Exp. Eye Res.*, *65*(4), 547–553.

[71] Jain-Vakkalagadda, B., Dey, S., Pal, D. & Mitra, A. K. (2003). Identification and functional characterization of a Na+-independent large neutral amino acid transporter, LAT1, in human and rabbit cornea. *Invest. Ophthalmol. Vis. Sci.*, *44*(7), 2919–2927.

[72] Nakauchi, T., Ando, A., Ueda-Yamada, M. et al., (2003). Prevention of ornithine cytotoxicity by nonpolar side chain amino acids in retinal pigment epithelial cells. *Invest. Ophthalmol. Vis. Sci.*, *44*(11), 5023–5028.

[73] Gandhi, M. D., Pal, D. & Mitra, A. K. (2004). Identification and functional characterization of a Na(+)-independent large neutral amino acid transporter (LAT2) on ARPE-19 cells. *Int. J. Pharm.*, *275*(1–2), 189–200.

[74] Anand, B. S. & Mitra, A. K. (2002). Mechanism of corneal permeation of Lvalyl ester of acyclovir: Targeting the oligopeptide transporter on the rabbit cornea. *Pharm. Res.*, *19*(8), 1194–1202.

[75] Xiang, C. D., Batugo, M., Gale, D. C., et al., (2009). Characterization of human corneal epithelial cell model as a surrogate for corneal permeability assessment: Metabolism and transport. *Drug Metab. Dispos.*, *37*(5), 992–998.

[76] Basu, S. K., Haworth, I. S., Bolger, M. B. & Lee, V. H. (1998). Proton-driven dipeptide uptake in primary cultured rabbit conjunctival epithelial cells. *Invest. Ophthalmol. Vis. Sci.*, *39*(12), 2365–2373.

[77] Berger, U. V. & Hediger, M. A. (1999). Distribution of peptide transporter PEPT2 mRNA in the rat nervous system. *Anat. Embryol.*, (Berl). *199*(5), 439–449.

[78] Macha, S. & Mitra, A. K. (2001). Ocular pharmacokinetics of cephalosporins using microdialysis. *J. Ocul. Pharmacol. Ther.*, *17*(5), 485–498.

[79] Talluri, R. S., Katragadda, S., Pal, D. & Mitra, A. K. (2006). Mechanism of Lascorbic acid uptake by rabbit corneal epithelial cells: Evidence for the involvement of sodium-dependent vitamin C transporter 2. *Curr. Eye Res.*, *31*(6),481–489.

[80] Janoria, K. G., Hariharan, S., Paturi, D., Pal, D. & Mitra, A. K. (2006). Biotin uptake by rabbit corneal epithelial cells: Role of sodium-dependent multivitamin transporter (SMVT). *Curr. Eye Res.*, *31*(10), 797–809.

[81] Hariharan, S., Janoria, K. G., Gunda, S., Zhu, X., Pal, D. & Mitra, A. K. (2006). Identification and functional expression of a carrier-mediated riboflavin transport system on rabbit corneal epithelium. *Curr. Eye Res.*, *31*(10), 811–824.

[82] Hosoya, K., Fujita, K. & Tachikawa, M. (2008). Involvement of reduced folate carrier 1 in the inner blood-retinal barrier transport of methyltetrahydrofolate. *Drug Metab. Pharmacokinet.*, *23*(4), 285–292.

[83] Mainardes, R. M., Urban, M. C., Cinto, P. O., et al. (2005). Colloidal carriers for ophthalmic drug delivery. *Curr. Drug Targets*, *6*(3), 363–371.

[84] Rabinovich-Guilatt, L., Couvreur, P., Lambert, G. & Dubernet, C. (2004). Cationic vectors in ocular drug delivery. *J. Drug Target*, *12*(9–10), 623–633.

[85] Schaeffer, H. E. & Krohn, D. L. (1982). Liposomes in topical drug delivery. *Invest. Ophthalmol. Vis. Sci.*, *22*(2), 220–227.

[86] Nagarsenker, M. S., Londhe, V. Y. & Nadkarni, G. D. (1999). Preparation and evaluation of liposomal formulations of tropicamide for ocular delivery. *Int. J. Pharm.*, *190*(1), 63–71.
[87] Hathout, R.M., Mansour, S., Mortada, N.D. & Guinedi, A.S. (2007). Liposomes as an ocular delivery system for acetazolamide: *In vitro* and *in vivo* studies. *AAPS Pharm. Sci. Tech.*, *8*(1), 1.
[88] Amrite, A. C. & Kompella, U. B. (2005). Size-dependent disposition of nanoparticles and microparticles following subconjunctival administration. *J. Pharm. Pharmacol.*, *57*(12), 1555–1563.
[89] Amrite, A. C., Edelhauser, H. F., Singh, S. R. & Kompella, U. B. (2008). Effect of circulation on the disposition and ocular tissue distribution of 20 nm nanoparticles after periocular administration. *Mol. Vis.*, *14*, 150–160.
[90] Cheruvu, N. P., Amrite, A. C. & Kompella, U. B. (2008). Effect of eye pigmentation on transscleral drug delivery. *Invest. Ophthalmol. Vis. Sci.*, *49*(1), 333–341.
[91] Kompella, U. B., Sundaram, S., Raghava, S. & Escobar, E. R. (2006). Luteinizing hormone-releasing hormone agonist and transferrin functionalizations enhance nanoparticle delivery in a novel bovine *ex vivo* eye model. *Mol. Vis.*, *12*, 1185–1198.
[92] Cheruvu, N. P., Amrite, A. C. & Kompella, U. B. (2009) Effect of diabetes on transscleral delivery of celecoxib. *Pharm. Res.*, *26*(2), 404–414.
[93] Peeters, L., Lentacker, I., Vandenbroucke, R. E., et al. (2008). Can ultrasound solve the transport barrier of the neural retina? *Pharm. Res.*, *25*(11), 2657–2665.
[94] Martin, N. E., Kim, J. W. & Abramson, D. H. (2008). Fibrin sealant for retinoblastoma: Where are we? *J. Ocul. Pharmacol. Ther.*, *24*(5), 433–438.
[95] Jiang, J., Gill, H. S., Ghate, D., et al. (2007). Coated microneedles for drug delivery to the eye. *Invest. Ophthalmol. Vis. Sci.*, *48*(9), 4038–4043.
[96] Jiang, J., Moore, J. S., Edelhauser, H. F. & Prausnitz, M. R. (2009). Intrascleral drug delivery to the eye using hollow microneedles. *Pharm. Res.*, *26*(2), 395–403.
[97] Zderic, V., Clark, J. I., Martin, R. W. & Vaezy, S. (2004). Ultrasound-enhanced transcorneal drug delivery. *Cornea*, *23*(8), 804–811.
[98] Vaka, S. R., Sammeta, S. M., Day, L. B. & Murthy, S. N. (2008). Transcorneal iontophoresis for delivery of ciprofloxacin hydrochloride. *Curr. Eye Res.*, *33*(8), 661–667.
[99] Frucht-Pery, J., Raiskup, F., Mechoulam, H., Shapiro, M., Eljarrat-Binstock, E. & Domb, A. (2006). Iontophoretic treatment of experimental pseudomonas keratitis in rabbit eyes using gentamicin-loaded hydrogels. *Cornea*, *25*(10), 1182–1186.
[100] Voigt, M., de Kozak, Y., Halhal, M., Courtois, Y. & Behar-Cohen, F. (2002). Down-regulation of NOSII gene expression by iontophoresis of anti-sense oligonucleotide in endotoxin-induced uveitis. *Biochem. Biophys. Res. Commun.*, *295*(2), 336–341.
[101] Raiskup-Wolf, F., Eljarrat-Binstock, E., Rehak, M., Domb, A. & Frucht-Pery, J. (2007). Transcorneal and transscleral iontophoresis of the dexamethasone phosphate into the rabbit eye. *Cesk Slov. Oftalmol.*, *63*(5), 360–368.
[102] Eljarrat-Binstock, E., Orucov, F., Frucht-Pery, J., Pe'er, J. & Domb, A. J. (2008a). Methylprednisolone delivery to the back of the eye using hydrogel iontophoresis. *J. Ocul. Pharmacol. Ther.*, *24*(3), 344–350.

[103] Eljarrat-Binstock, E., Domb, A. J., Orucov, F., Dagan, A., Frucht-Pery, J. & Pe'er, J. (2008b). *In vitro* and *in vivo* evaluation of carboplatin delivery to the eye using hydrogel-iontophoresis. *Curr. Eye Res.*, *33*(3), 269–275.

[104] Eljarrat-Binstock, E., Domb, A. J., Orucov, F., Frucht-Pery, J. & Pe'er, J. (2007). Methotrexate delivery to the eye using transscleral hydrogel iontophoresis. *Curr. Eye Res.*, *32*(7–8), 639–646.

[105] Hollo, G., Whitson, J. T., Faulkner, R., et al., (2006). Concentrations of betaxolol in ocular tissues of patients with glaucoma and normal monkeys after 1 month of topical ocular administration. *Invest. Ophthalmol. Vis. Sci.*, *47*(1), 235–240.

[106] Acheampong, A. A., Shackleton, M., John, B., Burke, J., Wheeler, L. & Tang-Liu, D. (2002). Distribution of brimonidine into anterior and posterior tissues of monkey, rabbit and rat eyes. *Drug Metab. Dispos.*, *30*(4), 421–429.

[107] Koevary, S. B. (2003). Pharmacokinetics of topical ocular drug delivery: Potential uses for the treatment of diseases of the posterior segment and beyond. *Curr. Drug Metab.*, *4*(3), 213–222.

[108] Koeberle, M. J., Hughes, P. M., Skellern, G. G. & Wilson, C. G. (2006). Pharmacokinetics and disposition of memantine in the arterially perfused bovine eye. *Pharm. Res.*, *23*(12), 2781–2798.

[109] Majumdar, S., Hingorani, T., Srirangam, R., Gadepalli, R. S., Rimoldi, J. M. & Repka, M. A. (2009). Transcorneal permeation of L- and Daspartate ester prodrugs of acyclovir: Delineation of passive diffusion versus transporter involvement. *Pharm. Res.*, *26* (5), 1261–1269.

[110] Anand, B. S., Katragadda, S., Nashed, Y. E. & Mitra, A. K. (2004). Amino acid prodrugs of acyclovir as possible antiviral agents against ocular HSV-1 infections: Interactions with the neutral and cationic amino acid transporter on the corneal epithelium. *Curr. Eye Res.*, *29*(2–3), 153–166.

[111] Gunda, S., Hariharan, S. & Mitra, A. K. (2006). Corneal absorption and anterior chamber pharmacokinetics of dipeptide monoester prodrugs of ganciclovir (GCV): *In vivo* comparative evaluation of these prodrugs with Val-GCV and GCV in rabbits. *J. Ocul. Pharmacol. Ther.*, *22*(6), 465–476.

[112] Majumdar, S., Nashed, Y. E., Patel, K., et al. (2005). Dipeptide monoester ganciclovir prodrugs for treating HSV-1-induced corneal epithelial and stromal keratitis: *In vitro* and *in vivo* evaluations. *J. Ocul. Pharmacol. Ther.*, *21*(6), 463–474.

*Chapter 7*

# TRANSDERMAL DELIVERY SYSTEMS

Intradermal delivery (IDD) has been and is being used as the route of choice for only a very limited number of vaccines, such as Bacille Calmette Guerin (BCG) for tuberculosis (TB) and in at least some countries for post-exposure rabies vaccination. It has also been investigated in recent decades as an alternative delivery route for several other vaccines, including hepatitis B (HBV), measles, and influenza.

Transdermal delivery systems to administer sufentanil through the skin are provided. The systems contain a sufficient amount of sufentanil to induce and maintain a constant state of analgesia when applied to a subject.

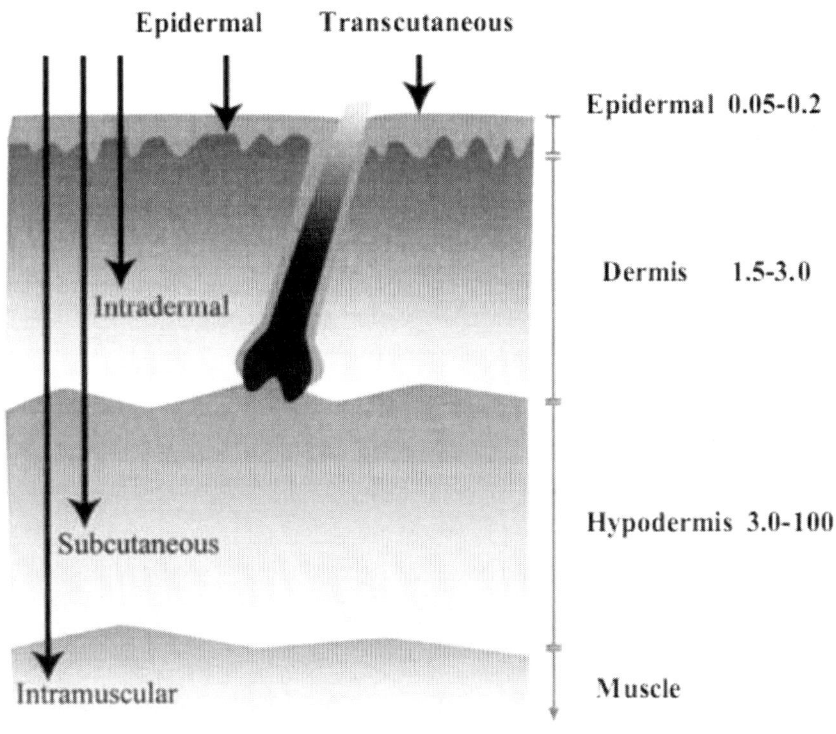


Figure 1. Schematic diagram that outlines the different layers that makeup the skin.

These systems are characterized as having one or more of the following features: a high control rate dosage form over the flux of sufentanil in the system; a high net flux of sufentanil in the system through the skin; lack of a permeation enhancer; an adhesive member that can demonstrate a superior shear time; a low coefficient variation in the net flux of sufentanil in the system; a high delivery efficiency, and a substantially constant state net flux of sufentanil in the system. Methods in using the transdermal delivery systems to administer a sufficient amount of sufentanil to induce and maintain analgesia for extended periods when applied to a subject are also provided. A *transdermal patch* is a medicated adhesive patch that is placed on the skin to deliver a specific dose of medication through the skin and into the bloodstream. Often, this promotes healing to an injured area of the body. There are advantages with using the transdermal drug delivery route over other types of medication delivery such as oral, topical, intravenous, intramuscular, etc. One convenience would be that the patch provides a controlled release of the medication into the patient, usually through either a porous membrane covering a reservoir of medication or through body heat melting thin layers of medication embedded in the adhesive. The main disadvantage to transdermal delivery systems stems from the fact that the skin is a very effective barrier; as a result, only medications whose molecules are small enough to penetrate the skin can be delivered by this method. A wide variety of pharmaceuticals are now available in transdermal patch form. Before these patches go into the market, they have to be carefully studied. One way to study these patches are through the use of Franz Diffusion Cell systems. This system is used to study the temperature effects on a permeated amount for a specific drug on a certain type of membrane, which in this case would be the membrane used in these patches. A Franz Diffusion Cell system is composed of a receptor and a donor cell. In many of these research studies the following procedure is used. The donor cell is set at a specific temperature (the temperature of the environment), while the receptor cell is set at different one (temperature of the body). Different runs are performed using different temperatures to study the impact of temperature on the release of a certain medicament through a certain type of membrane. Although different concentrations of the medicament are used in this study, they do not affect the amount permeated through the membrane (the process is constant). From chemical kinetics it can be concluded that these studies are zero order, since the concentration plays no role in the permeated amount through the membrane. Some pharmaceuticals must be combined with substances, such as alcohol, within the patch to increase their ability to penetrate the skin to be used as a transdermal patch. Others can overwhelm the body if applied in only one place, and are often cut into sections and applied to different parts of the body to avoid this, such as nitroglycerin. Many molecules, however, such as insulin, are too large to pass through the skin without it being modified in some way. Several new technologies are being investigated to allow larger molecules to be delivered transdermally.

Transdermal drug delivery system has been in existence for a long time. In the past, the most commonly applied systems were topically applied creams and ointments for dermatological disorders. The occurrence of systemic side effects with some of these formulations is indicative of absorption through the skin. A number of drugs have been applied to the skin for systemic treatment. In a broad sense, the term transdermal delivery system includes all topically administered drug formulations intended to deliver the active ingredient into the general circulation. Transdermal therapeutic systems have been designed to provide controlled continuous delivery of drugs via the skin to the systemic circulation. Moreover, it overcomes various side effects like painful delivery of the drugs and the first pass metabolism of the drug occurred by other means of drug delivery systems. So, this transdermal drug delivery system has currently been a great field of interest. Many drugs which can be injected directly into the blood stream via the skin have been formulated. The main advantages of this system are that there is a controlled release of the drug and the medication is painless. The drug is mainly delivered to the skin with the help of a transdermal patch which adheres to the skin. A transdermal patch has several components like liners, adherents, drug reservoir, drug release membrane, etc. All these features play a vital role in the release of the drug via the skin. Various types of patches along with various methods of applications have been invented to deliver the drug when using the transdermal patch. Because the transdermal patch has such advantages, it has become one of the highly researched fields among the various drug delivery systems.

## Types of Transdermal Patches

### Multi-Layer Drug-in-Adhesive

The multi-layer drug-in-adhesive patch is similar to the single-layer system in that both adhesive layers are also responsible for drug release. The multi-layer system is different, however, that it adds another layer of drug-in-adhesive, usually separated by a membrane (but not in all cases). This patch also has a temporary liner-layer and a permanent backing.

### Reservoir

Unlike the single-layer and multi-layer drug-in-adhesive systems, the reservoir transdermal system actually has a separate drug layer. The drug layer consists of a liquid compartment which contains the drug solution or suspension that is separated by a adhesive layer. This patch is also protected by a backing layer. In this type of system the rate of release is of zero order.

### Matrix

The matrix system has a drug layer that is a semi-solid matrix to contain the drug solution or suspension. The adhesive layer in this patch surrounds the drug layer partially overlaying it. The new hydrogel patches are a good example of a semi-solid matrix system.

### Vapor Patch

In this type of patch the adhesive layer not only serves to adhere to the various layers together but can also release a vapor. The vapor patches are relatively new on the market and can release essential oils for up to 6 hours. The vapors patches release essential oils and are used mainly in cases of decongestion. Other vapor patches on the market are controller vapor patches that improve the quality of sleep. Vapor patches that reduce the quantity of cigarettes that one smokes in a month are also available on the market. There is a range of medication that use patches, such as: Nicotine: to quit tobacco smoking, Fentanyl: analgesic for severe pain, Estrogen: menopause and osteoporosis, Nitroglycerin: angina, Lidocaine: peripheral pain of shingles (herpes zoster). Recent developments expanded their use to the delivery of hormonal contraceptives, antidepressants and even pain killers and stimulants for Attention Deficit Hyperactivity Disorder/ADHD. The adhesive layer contains the drug. The adhesive layer not only serves to adhere the various layers together, it is also responsible for the releasing the drug into the skin. The adhesive layer is surrounded by a temporary liner and a backing. Another novel type, which is also similar to the single layer, contains an immediate drug release layer and then another layer that is for a controlled release along with the adhesive layer. The adhesive layer is responsible for the release of the drug. This patch also has a temporary liner-layer and a permanent backing. Vapor patches available on the market are frequently used to improve the quality of sleep and reduce the urge of cigarette smoking. A system of the drug reservoir is embedded between an impervious backing layer and a rate-controlling membrane. The drug releases only through the rate-controlling membrane, which can be micro-porous or nonporous. In the drug reservoir compartment, the drug can be in the form of a solution, suspension, gel, or dispersed in a solid polymer matrix. Hypoallergenic adhesive polymer can be applied as outer surface polymeric membrane which is compatible with the drug. The drug reservoir is formed by dispersing the drug into an adhesive polymer and then spreading the medicated adhesive polymer by solvent casting or melting (in the case of hot-melt adhesives) on an impervious backing layer. On top of the reservoir, unmediated adhesive polymer layers are applied for protection purpose. There exists another type of patch where the drug is dispersed homogenously in a hydrophilic or lipophilic polymer matrix. This drug containing polymer is in the form of a disc that is fixed onto an occlusive base plate within a compartment fabricated by a drug impermeable backing layer. Instead of applying the adhesive on the face of the drug reservoir, it is spread along the circumference to form a strip on the adhesive rim. The drug delivery system is a combination of a reservoir and a matrix-dispersion system. The drug reservoir is formed by first suspending the drug in an aqueous solution of water soluble polymer and then dispersing the solution homogeneously in a lipophilic polymer to form thousands of unreachable, microscopic spheres of drug reservoirs. This thermodynamically unstable dispersion is stabilized quickly by immediately cross-linking the polymer *in situ* by using cross-linking agents. A prototype patch can be fabricated for this heat sealable polyester film (type 1009, 3m) with a concave of 1cm diameter as the backing membrane. A drug sample is dispensed into the concave membrane, covered by a TPX {poly(4-methyl-1-pentene)}asymmetric membrane, and sealed by an adhesive. These are fabricated by using the dry/wet inversion process. TPX is dissolved in a mixture of solvent (cyclohexane) and nonsolvent additives at 60°c to form a polymer solution. The polymer solution is kept at 40°C for 24 hrs and cast on a glass plate to a pre-determined thickness with a gardner knife.

After that the casting film is evaporated at 50°C for 30 sec, then the glass plate is to be immersed immediately in coagulation bath [maintained the temperature at 25°C]. After 10 minutes of immersion, the membrane can be removed, air dry in a circulation oven at 50°C for 12 hrs].

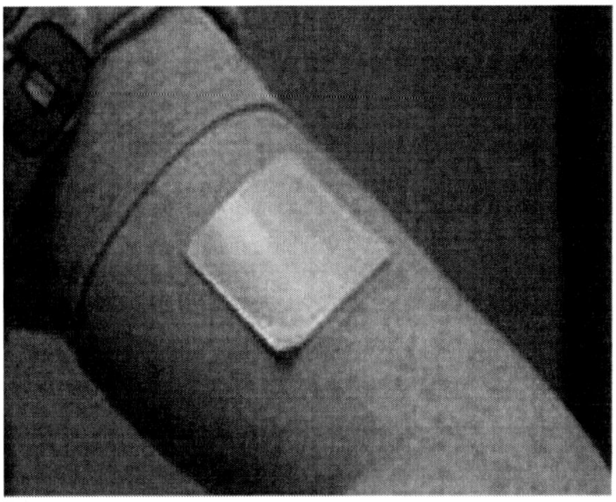

Figure 2. Single-layer Drug-in-Adhesive patch.

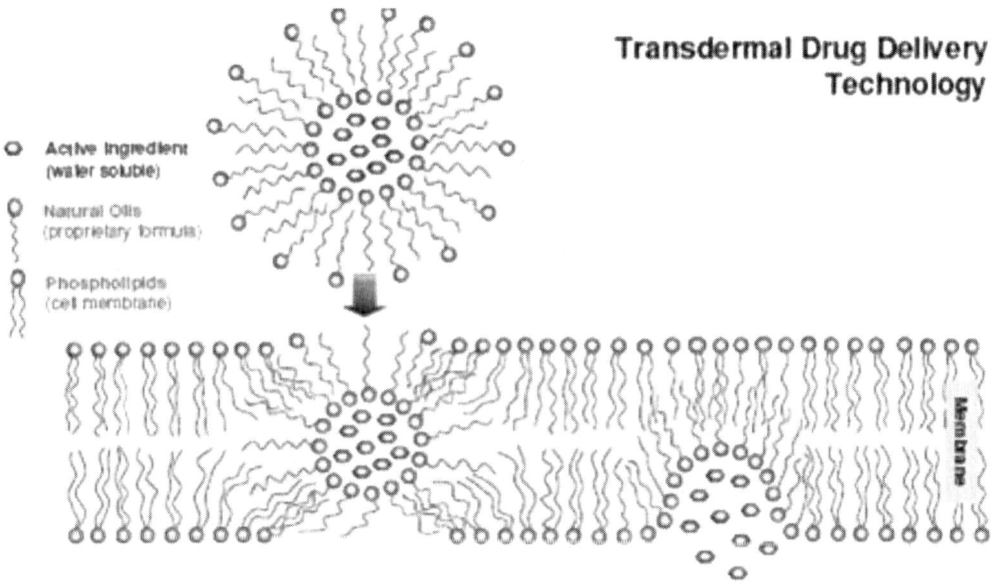


Solutions containing polymers in various ratios are used in an organic solvent. A calculated amount of drug is dissolved in half the quantity of same organic solvent. Enhancers in different concentrations are dissolved in the other half of the organic solvent and then added. Di-N-butylphthalate is added as a plasticizer into drug polymer solution. The total contents are to be stirred for 12 hrs and then poured into a circular teflon mold. The molds are to be placed on a level surface and covered with an inverted funnel to control solvent vaporization under a laminar flow hood model with an air speed of 0.5 m/s. The solvent is allowed to evaporate for 24 hrs.

The dried films are stored for another 24 hrs at 25±0.5°C in a desiccators containing silica gel before evaluation to eliminate aging effects. The type films are evaluated within one week of their preparation.

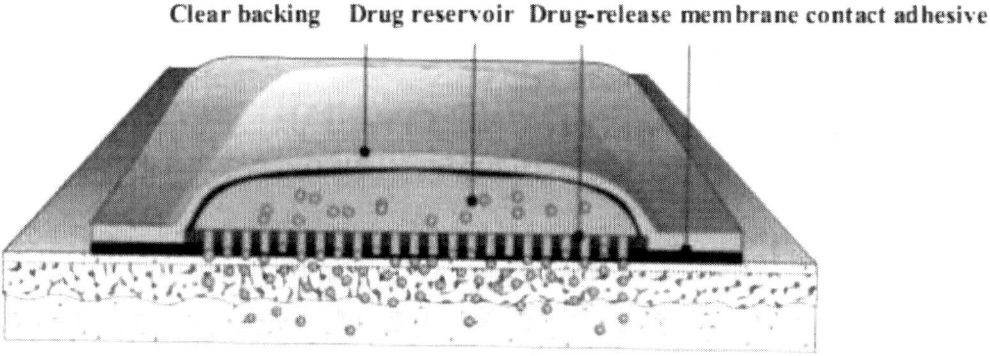


Transdermal drug delivery system may produce unstable matrices if the loading dose is greater than 10 mg. Aluminium backed adhesive film method is a suitable one. For a similar preparation, chloroform was the choice of solvent because most of the drugs, as well as adhesive are soluble in chloroform. The drug is dissolved in chloroform and adhesive material is added to the drug solution and dissolved. A custom made aluminium former is lined with aluminium foil and the ends blanked off with tightly fitting cork blocks. The drug is then dissolved in a polymer solution along with a plasticizer. This solution is then stirred for 10-15 minutes to produce a homogenous dispersion and poured in to a level mercury surface, covered by an inverted funnel to control solvent evaporation. In this method, the drug is dispersed in a mixture of water and propylene glycol containing a carbomer 940 polymer and stirred for 12 hrs in magnetic stirrer. The dispersion is neutralized and made viscous by the addition of triethanolamine. Buffer pH 7.4 can be used in order to obtain solution gel, if the drug solubility in aqueous solution is very poor. The formed gel will be incorporated in the IPM membrane.

In order to prepare the target transdermal therapeutic system, 1% carbopol reservoir gel, polyethelene (PE), ethylene vinyl acetate copolymer (EVAC) membranes can be used as the rate control membranes. If the drug is not soluble in water, propylene glycol is used for the preparation of the gel. A drug is dissolved in propylene glycol, where carbopol resin is added to this solution and neutralized by using 5% w/w sodium hydroxide solution. The drug (in gel form) is placed on a sheet of backing layer covering the specified area. A rate controlling membrane is placed over the gel and the edges are sealed by heat to obtain a leak proof device. The proliposomes are prepared by the carrier method which uses a film deposition technique. From the earlier reference, the drug and lecithin in the ratio of 0.1:2.0 can be used as an optimized one. The proliposomes are prepared by taking 5mg of mannitol powder in a 100 ml round bottom flask which is kept at 60-70°c temperature. The flask is rotated at 80-90 rpm and to complete the process to dry the mannitol under vacuum for 30 minutes. After drying, the temperature of the water bath is adjusted to 20-30°C. Drug and lecithin are dissolved in a suitable organic solvent mixture; a 0.5ml aliquot of the organic solution is introduced into the round bottomed flask at 37°C. Then, after complete drying, the second aliquot (0.5ml) of the solution is added. After the last loading, the flask containing proliposomes are connected in a lyophilizer and subsequently the drug loaded mannitol powders (proliposomes) are placed in a desiccator over night and then filtered through a 100 mesh. The collected powder is transferred into a glass bottle and stored at a freezing temperature.

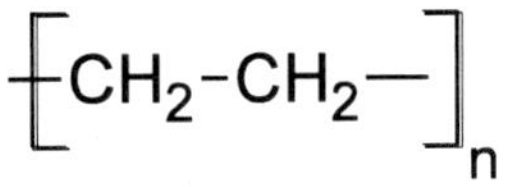

Poly(ethylene)

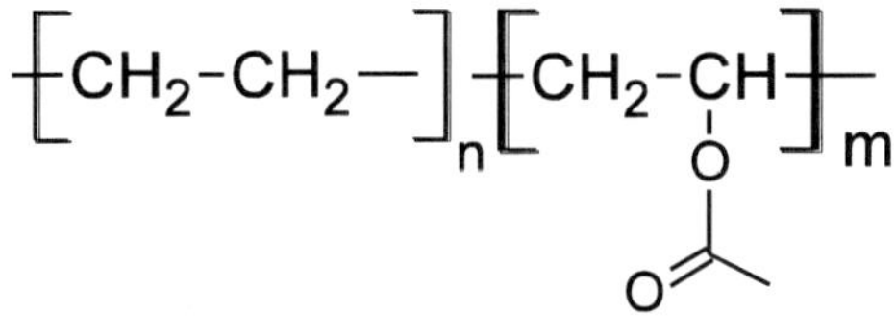

Polyethylene Vinyl Acetate

To prepare a free film of cellulose acetate the process is done by casting it on a mercury surface. A polymer solution of 2% w/w is prepared by using chloroform. Plasticizers are incorporated at a concentration of 40% w/w of its polymer weight. Five ml of polymer solution is poured in a glass ring and placed over the mercury surface in a glass petri dish. The rate of evaporation of the solvent is controlled by placing an inverted funnel over the petri dish. The film formation is noted by observing the mercury surface after complete evaporation of the solvent. The dry film is separated out and stored between the sheets of wax paper in a desiccator until use. Free films of different thicknesses can be prepared by changing the volume of the polymer solution.

Excipients are integral components for almost all pharmaceutical dosage forms. The stability of a formulation amongst other factors depends on the compatibility of the drug with the excipients. The drug and the excipients must be compatible with one another to produce a product that is stable, thus it is mandatory to detect any possible physical or chemical interaction as it can affect the bioavailability and stability of the drug. If the excipients are new and have not been used in formulations containing the active substance, the compatibility studies play an important role in formulation development. Interaction studies are commonly carried out in Thermal analysis, FT-IR, UV and chromatographic techniques by comparing their physicochemical characters such as assay, melting endotherms, characteristic wave numbers, absorption maxima, etc.

The thickness of the drug loaded patch is measured at different points by using a digital micrometer and will determine the average thickness and standard deviation for each patch to ensure the same thickness has been prepared for every patch. The prepared patches are then dried at 60°c for 4hrs before testing. A specified area of patch is cut in different areas of the patch and then weighed by a digital balance. The average weight and standard deviation values are calculated from the individual weights. A strip is specifically cut evenly then repeatedly folded at the same place to test its limitations. The number of times the film is folded at the same place without breaking provides the value for its folding endurance.

An accurately weighed portion of the patch is cut into small pieces and transferred to a specific volume volumetric flask, and then dissolved into a suitable solvent and sonicate for complete drug extraction. Then this resulting solution was allowed to settle for about an hour, and the supernatant was suitably diluted to give the desired concentration with suitable solvent. The solution was filtered using 0.2μm membrane filter and analyzed by suitable analytical technique (UV or HPLC) and the drug content per piece is then calculated. A test is then performed to examine the drug crystals on the patch by a polariscope. A specific surface area of the piece is kept on the object slide and observed for drug crystals to distinguish whether the drug is present as a crystalline form or amorphous form in the patch.

A new test is performed to measure the cohesive strength of the adhesive polymer. The adhesive can be influenced by molecular weight, the degree of cross-linking and the composition of the polymer, type and the amount of tackifier added. An adhesive coated tape is applied onto a stainless steel plate; a specified weight is hung from the tape, to affect it by pulling it in the direction parallel to the plate. Shear adhesion strength is determined by measuring the time it takes to pull the tape off the plate. The longer it takes for its removal, the greater the shear strength is. The force required to remove an adhesive coating form from the test substrate is referred to as peel adhesion. Molecular weight of adhesive polymer, the type and amount of additives are the variables that will determine the peel adhesion properties. A single tape is applied to a stainless steel plate or a backing membrane of choice and then the tape is pulled from the substrate at a 180° angle, and the force required to remove the tape is measured. It is a qualitative test applied for obtaining the tack property determination of the adhesive. The thumb is simply pressed on the adhesive and the relative tack property is detected. Three longitudinal strips are cut from each film to monitor a different portion. For instance, like one strip from the center of the patch, then another one from its left side, and another strip taken from its right side. The length of each strip is then measured to verify the variation in length because of non-uniformity in flatness is measured by determining percent of constriction, with 0% constriction equivalent to 100% flatness.

The tape is pulled away from the substrate at 90°C at a speed of 12 inches/min. The peel force that is required to break the bond between adhesive and substrate is measured and recorded as a tack value, and expressed in ounces or grams per inch width. In this test, the tip of a clean probe with a defined surface roughness is brought into contact with adhesive, and when a bond is formed between probe and adhesive. The subsequent removal of the probe mechanically breaks it. The force required to pull the probe away from the adhesive at a fixed rate is recorded as the tack and it is expressed in grams. The paddle over disc method (USP apparatus V) can be employed for assessment of the release of the drug from the prepared patches. Dry films of known thickness is cut into a definite shape, weighed, and fixed over a glass plate with an adhesive. The glass plate is then placed in a 500-mL of the dissolution medium or phosphate buffer (pH 7.4), and the apparatus is equilibrated to 32± 0.5°C. The paddle is then set at a distance of 2.5 cm from the glass plate and operated at a speed of 50 rpm. Samples (5- mL aliquots) were withdrawn at appropriate time intervals up to 24 h and analyzed by UV spectrophotometer or HPLC. This experiment is performed in triplicate and the mean value results are then calculated.

An *in vitro* permeation study was carried out by using a diffusion cell method on the full thickness of the abdominal skin of male Wistar rats weighing 200 to 250g. Hair from the abdominal region was removed carefully by using a electric clipper; the dermal side of the skin was thoroughly cleaned with distilled water to remove any adhering tissues or blood vessels, equilibrated for an hour in dissolution medium or phosphate buffer pH 7.4 before starting the experiment and was placed on a magnetic stirrer with a small magnetic needle for uniform distribution of the diffusant. The temperature of the cell was maintained at 32 ± 0.5°C using a thermostatically controlled heater. The isolated rat skin piece was mounted between the compartments of the diffusion cell, with the epidermis facing upward into the donor compartment. Sample volume of the definite volume was then removed from the receptor compartment at regular intervals, and an equal volume of fresh medium was replaced. Samples were filtered through a filtering medium and analyzed spectrophotometrically or HPLC. Flux results could then be determined directly as the slope of the curve between the steady-state values of the amount of drug permeated (mg $cm^{-2}$) vs.

time in hours and permeability coefficients were deduced by dividing the flux by the initial drug load (mg $cm^{-2}$).

Skin irritation and sensitization testing were then performed on healthy rabbits (average weight 1.2 to 1.5 kg). The dorsal surface (50$cm^2$) of the rabbit were cleaned and hair removed from the clean dorsal surface by shaving and cleaning the surface using rectified spirit and the representative formulations were applied over the skin. The patch was removed after 24 hrs and the skin conditions observed and classified into 5 grades on the basis of the severity of skin injury. Stability studies were conducted according to the ICH guidelines by storing the TDDS samples at 40±0.5°c and 75±5% RH for 6 months. The samples were withdrawn at 0, 30, 60, 90, and 180 days and analyzed suitably for drug content. The development of transdermal drug delivery systems is a multi-disciplinary activity that encompasses, fundamental feasibility studies that start from the selection of a drug molecule to the demonstration of sufficient drug flux in an *ex vivo* and/or *in vivo* model. The fabrication of a drug delivery system must meet all the stringent needs that are specific to the drug molecule (physicochemical and stability factors), the patient (comfort and cosmetic appeal), the manufacturer (scale-up and manufacturability), and most important, the economy.

Polymers are the backbone of a transdermal drug delivery system. Systems for transdermal delivery are fabricated as multilayered polymeric laminates in which a drug reservoir or a drug–polymer matrix is sandwiched between two polymeric layers: an outer impervious backing layer that prevents the loss of drug through the backing surface and an inner polymeric layer that functions as an adhesive and/or rate-controlling membrane. Transdermal drug delivery systems are broadly classified into three types, the reservoir, the matrix, and the vapor.

The drug reservoir is embedded between an impervious backing layer and a rate-controlling membrane. The drug releases only through the rate-controlling membrane, which can be microporous or nonporous. In the drug reservoir compartment, the drug can be in the form of a solution, suspension, or gel or dispersed in a solid polymer matrix. On the outer surface of the polymeric membrane a thin layer of drug-compatible, hypoallergenic adhesive polymer can be applied. Another novel drug delivery system is the combination of the reservoir and matrix-dispersion system. The drug reservoir is formed by first suspending the drug in an aqueous solution of water-soluble polymer and then dispersing the solution homogeneously in a lipophilic polymer to form thousands of unleachable, microscopic spheres of drug reservoirs. The thermodynamically unstable dispersion is stabilized quickly by immediately cross-linking the polymer *in situ*. Transdermal drug delivery technology represents one of the most rapidly advancing areas for drug delivery. This growth is catalyzed by developments in the field of polymer science. This article focuses on the polymeric materials used in transdermal delivery systems, with emphasis on the materials' physicochemical and mechanical properties, and it seeks to guide formulators in the selection of polymers. Polymers are used in transdermal delivery systems in various ways, to include matrix formers, rate-controlling membranes, pressure-sensitive adhesives (PSAs), backing layers, and release liners. The drug reservoir is formed by dispersing the drug into an adhesive polymer and then spreading the medicated polymer adhesive by solvent casting or by melting the adhesive (in the case of hot-melt adhesives) onto an impervious backing layer. On top of the reservoir, layers of unmedicated adhesive polymer are applied. The drug is dispersed homogeneously in a hydrophilic or lipophilic polymer matrix. This drug containing polymer disc then is fixed onto an occlusive base plate in a compartment fabricated with a drug-impermeable backing layer. Instead of applying the adhesive on the face of the drug reservoir, it is spread along the circumference to form a strip on the adhesive rim.

Polymers used in transdermal delivery systems should have biocompatibility and chemical compatibility with the drug and other components of the system such as penetration enhancers and PSAs. They also should provide consistent, effective delivery of a drug throughout the product's intended shelf-life or delivery period and have generally – recognized-as-safe status. From an economic point of view, a delivery tool kit rather than a single delivery tool is most effective. Companies involved in the field of transdermal delivery concentrate on a few selective polymeric systems. For example, Alza Corporation (Mountain View, CA) mainly concentrates on ethylene vinyl acetate (EVA) copolymers or microporous polypropylene, and Searle Pharmacia (Barceloneta, PR) concentrates on silicone rubber. A review of the marketed transdermal products and the formulations that are reported in various research publications reveals an enormous diversity of polymers used in the formulation, engineering, and manufacture of drug products.

Polymer selection and design must be considered when striving to meet the diverse criteria for the fabrication of effective transdermal delivery systems. The main challenge is in the design of a polymer matrix, followed by optimization of the drug loaded matrix not only in terms of release properties, but also with respect to its adhesion–cohesion balance, physicochemical properties, and compatibility and stability with other components of the system as well as with skin. A monolithic solid-state design often is preferred for passive transdermal delivery systems because of manufacturing considerations and cosmetic appeal. Biocompatibility of PEGs makes them the polymers of choice for numerous biomedical applications. Proteins can be delivered by PEGs cross-linked with tris(6-isocyanatohexyl) isocyanurate by means of a urethane–allophanate bond to obtain polymer networks capable of swelling in phosphate-buffered saline or ethanol and forming gels. These systems have been shown to release the solutes in a biphasic manner.

$$\left[ CH_2-CH_2-O \right]_n -CH_2-CH_2-O-$$

Poly(ethylene glycol)

Acrylic-acid matrices with plasticizers have been used to make drug–polymer matrix films for transdermal delivery systems. Some of the polymers that have been reported are Eudragit RL PM, Eudragit S-100, Eudragit RS PM, and Eudragit E-100 (Röhm America, Piscataway, NJ). Eudragit NE-40D (a copolymer of ethyl acrylate and methyl methacrylate), a nonadhesive hydrophobic polymer, also has been used as a matrix former. The release rates of drugs from these matrix systems are more closely described by the square-root-of-time model.

EC and PVP matrix films with 30% dibutyl phthalate as a plasticizer have been fabricated to deliver diltiazem hydrochloride and indomethacin. The addition of hydrophilic components such as PVP to an insoluble film former such as ethyl cellulose tends to enhance its release-rate constants. This outcome can be attributed to the leaching of the soluble component, which leads to the formation of pores and thus a decrease in the mean diffusion path length of drug molecules to release into the dissolution medium. The result is higher dissolution rates. Substances such as PVP act as anti-nucleating agents that retard the crystallization of a drug. Thus, they play a significant role in improving the solubility of a drug in the matrix by sustaining the drug in an amorphous form so that it undergoes rapid solubilization by penetration of the dissolution medium.

## REFERENCES


[1] Chien, Y. W. (1987). Transdermal therapeutic systems. In *Controlled Drug Delivery: Fundamentals and Applications*. J.R. Robinson and V.H.L. Lee, (eds.). Marcel Dekker, Inc., New York, NY, 2d edit., 523–552.

[2] Davis, S. S. & Illum, L. (1989). Drug delivery systems for challenging molecules, *Int. J. Pharm.*, *176*, 1–8.

[3] Baker, R. W. & Heller, J. Material selection for transdermal delivery systems. In: *Transdermal Drug Delivery: Developmental Issues and Research Initiatives*, J. Hadgraft and R.H. Guys, (eds.), Marcel Dekker, Inc., New York,, 293–311.

[4] Wolff, H.-M. (2000). Optimal process design for the manufacturing of transdermal drug delivery systems, *PSTT*, *3*(5), 173–181.

[5] L. Bromberg, Cross-linked poly(ethylene glycol) networks as reservoirs for protein delivery, *J. Appl. Poly. Sci.*, *59*, 459–466 (1996).

[6] Costa, P., et al. (1997). Design and evaluation of a lorazepam transdermal delivery system, *Drug Dev. Ind. Pharm.*, *23*(10), 939–944.

[7] Minghetti, P., et al. (1999). Dermal patches for the controlled release of miconazole: Influence of the drug concentration on the technical characteristics, *Drug Dev. Ind. Pharm.*, *25*, 679–684.

[8] Ramarao, P. & Diwan, P. V. (1998). Formulation and *in vitro* evaluation of polymeric films of diltiazem hydrochloride and indomethacin for transdermal administration, *Drug Dev. Ind. Pharm.*, *24* (4), 327–336.

[9] Guyot, M. & Fawaz, F. (2000). Design and *in vitro* evaluation of adhesive matrix for transdermal delivery of propranolol, *Int. J. Pharm.*, *204*, 171–182.

[10] Florence, A. T. & Attwood, D. (1982). *Physico-Chemical Principles of Pharmacy*, Chapman and Hall, New York, p. 206.

[11] Walde, P., et al. (1990). Lecithin microemulsion gels as a matrix for transdermal delivery of drugs, *International Symposium on Controlled Release of Bioactive Materials*, Controlled Release Society, Inc., Minneapolis, MN, 421–422.

[12] Bhatnagar, S. & Vyas, S. P. (1994). Organogel-based system for transdermal delivery of propranolol, *J. Microencapsulation*, *11*(4), 431–438.

[13] Willimann, H., et al. (1992). Lecithin organogel as matrix for transdermal transport of drugs, *J. Pharm. Sci.*, *81*(9), 871–874.

[14] Gale, R. & Spitze, L. A. (1981). Permeability of camphor in ethylene-vinyl acetate copolymers, *Proceedings: Eighth International Symposium on Controlled Release of Bioactive Materials*, Controlled Release Society, Minneapolis, MN, p. 183.

[15] Boretos, J. W., Detmer, D. E. & Donachy, J. H. (1971). Segmented polyurethane: A polyether polymer, II: Two years' experience, *J. Biomed. Mat. Res.*, *5*, 373.

[16] Kambe, T. N., et al. (1999). Microbial degradation of polyurethane, polyester polyurethanes and polyether polyurethanes, *Appl. Microbiol. Biotechnol.*, *51*, 134–140.

[17] Lyman, D. J. & Loo, B. H. (1967). New synthetic membranes for dialysis, IV: A copolyether-urethane membrane system, *J. Biomed. Mater. Res.*, 1(17).

[18] Baker, R. W., et al. Development of an estriol-releasing intrauterine device, *J. Pharm. Sci.*, *68* (1), 20.

[19] Pocius, A. V. (1991). Adhesives. In: *Kirk-Othmer Encyclopedia of Chemical Technology*, M. Howe-Grants, (ed.), Wiley-Interscience, New York, 4th edit., 445–466.

[20] Franz, T. J., et al. (1991). Transdermal Delivery. In: *Treatise on Controlled Drug Delivery: Fundamentals, Optimization, Applications*, A. Kydonieus, (ed.), Marcel Dekker, Inc., New York, 341–421.

[21] Barnhart, S. (1998). Critical role of PSAs in transdermal drug delivery, *Adhesives and Sealants Industry*, April 1-6, 1998.

[22] Tan, H. S. & Pfister, W. R. (1999). Pressure-sensitive adhesives for transdermal drug delivery systems, *PSTT 2*, (2), 60–69.

[23] Pfister, W. R. & Hsieh, D. S. T. (1990). Permeation enhancers compatible with transdermal drug delivery systems, Part II: System design considerations, *Pharm. Technol.*, *14*(10), 54–60.

[24] Pfister, W. R. (1989). Customizing silicone adhesives for transdermal drug delivery systems, *Pharm. Technol.*, *13* (3), 126–138.

[25] Fuller, C. S., Frosch, C. J. & Pape, N. R. (1940). X-ray examination of polyisobutylene, *J. Am. Chem. Soc.*, *62*, 1905–1913.

[26] Wood, L. A. (1976). *Rubber Chem. Technol.*, *49*, 189.

[27] Kruge, E. & Wand, H. C. (1991). Butyl rubber. In: *Kirk-Othmer Encyclopedia of Chemical Technology*, M. Howe-Grants, (ed.), Wiley-Interscience, New York, 4th edit., 934–955.

[28] Chen, Y.-L., et al. (1986). Transdermal Therapeutic Systems for the Administration of Naloxone, Naltrexone and Nalbuphine, US Patent No. 4,573,995 (assigned to Alza Corporation, Mountain View, CA, ).

[29] Takayasu, T., et al. "Patch," US Patent No. 6,211,425 (assigned to Saitama Daiichi Seiyaku Kabushiki Kaisha, Kasukabe, JP).

[30] Novak, R. W. (1991). Acrylic ester polymers. In: *Kirk-Othmer Encyclopedia of Chemical Technology*, M. Howe-Grants, (ed.), Wiley-Interscience, New York, 4th ed., 314–343.

[31] Shetter, J. A. (1963). *J. Polym. Sci. Part B.*, *1*, 209.

[32] Burgess, A. R. (1952). *Chem. Ind.*, *78*.

[33] Woodard, J. T. & Metevia, V. L. (1987). Transdermal Drug Delivery Devices with Amine-Resistant Silicone Adhesives, US Patent No. 4,655,767 (assigned to Dow Corning Corporation, Midland, MI).

[34] Muraoka, T., et al. (2000). Percutaneous Absorption Preparation, US Patent No. 6,132,761 (assigned to Nitto Denko Corporation, Osaka, JP).

[35] Pretzer, P. W. & Sweet, R. P. (1999).Silicone Pressure-Sensitive Adhesive Composition Containing Functionalized Polyisobutylene, US Patent No. 5,939,477 (assigned to Dow Corning Corporation, Midland, MI).

[36] Godbey, K. J. (1996). Improving patient comfort with nonocclusive transdermal backings. In: *American Association of Pharmaceutical Scientists*, 1–2.

[37] Rolf, D., Sjoblom, U. & Elisabeth, K. (2000). Method of Forming Adhesive Patch for Applying Medication to the Skin, U.S. Patent No. 6,096,333 (assigned to LecTec Corporation, Minnetonka, MN).
[38] Santoro, A., et al. (2000). Pharmaceutical development and characteristics of a new glyceryl trinitrate patch, *Arzneimittel-Forschung/Drug Res.*, *50* (2), 897–903.
[39] *Polymer Handbook*, Interscience Publishers, New York, 1975.
[40] Gardon, J. L. (1965). *Encyclopedia of Polymer Science and Technology*, N.M. Bikaless, (ed.), Interscience Publishers, New York, 833–862.
[41] Gabiga, H., Cal, K. & Janicki, S. (2000). Effect of penetration enhancers on isosorbide dinitrate penetration through rat skin from a transdermal therapeutic system, *Int. J. Pharm.*, *199*, 1–6.
[42] Maillard-Salin, D. G., Becourt, P. & Couarraze, G. (2000a). Physical evaluation of a new patch made of a progestomimetic in a silicone matrix, *Int. J. Pharm.*, *199*, 29–38.
[43] Maillard-Salin, D. G., Becourt, P. & Couarraze, G. (2000b). A study of the adhesive-skin interface: Correlation between adhesion and passage of a drug, *Int. J. Pharm.*, *200*, 121–126.
[44] El-Kattan, A. F., Asbil, C. S. & Michniak, B. B. (2000). The effect of terpene enhancer lipophilicity on the percutaneous permeation of hydrocortisone formulated in HPMC gel systems, *Int. J. Pharm.*, *198*, 179–189 .
[45] Minghetti, P., et al. (2000). Development of local patches containing melilot extract and *ex vivo–in vivo* evaluation of skin permeation, *Eur. J. Pharm. Sci.*, *10*, 111–117.
[46] Sutinen, R., Paronen, P. & Urtti, A. (2000). Water-activated, pH-controlled patch in transdermal administration of timolol I: Preclinical tests, *Eur. J. Pharm. Sci.*, *11*, 19–24.
[47] Iwase, H., et al. (2000). Transdermal absorption of L-dopa from a new system composed of two separate layers of L-dopa and hydrogel in rats, *Drug Dev. Ind. Pharm.*, *26*(7), 755–759.
[48] Pongjanyakul, T., Prakongpan, S. & Priprem, A. (2000). Permeation studies comparing cobra skin with human skin using nicotine transdermal patches, *Drug Dev. Ind. Pharm.*, *26*(6): 635–642.
[49] Ohmori, S., et al. (2000). Transdermal delivery of the potent analgesic dihydroetorphine: Kinetic analysis of skin permeation and analgesic effect in the hairless rat, *J. Pharm. Pharmacol.*, *52*, 1437–1449.
[50] Cho, Y.-J. & Choi, H.-K. (1998). Enhancement of percutaneous absorption of ketoprofen: Effect of vehicles and adhesive matrix, *Int. J. Pharm.*, *169*, 95–104.
[51] Kim, J.-H., Cho, Y.-J. & Choi, H.-K. (2000). Effect of vehicles and pressure-sensitive adhesives on the permeation of tacrine across hairless mouse skin, *Int. J. Pharm.*, *196*, 105–113.
[52] Ebert, C. D., et al. (1987). Development of a novel transdermal system design, *J. Controlled Rel.*, *6*, 107–111.
[53] Kokubo, T., Sugibayashi, K. & Morimoto, Y. (1991). Diffusion of drug in acrylic-type pressure-sensitive adhesive matrices on the drug diffusion, *J. Controlled Rel.*, *17*, 69–78.
[54] Okano, T., et al. (1987). Control of drug concentration-time profiles *in vivo* by zero-order transdermal delivery systems, *J. Controlled Rel.*, *6*, 99–106.

[55] Keshary, P. R., Huang, Y. C. & Chien, Y. W. (1985). Mechanism of transdermal controlled nitroglycerin administration III: Control of skin- permeation rate and optimization, *Drug Dev. Ind. Pharm.*, *11* (6,7), 1213–1253.
[56] Roy, S. D., et al. (1996). Controlled transdermal delivery of fentanyl: Characterizations of pressure-sensitive adhesives for matrix patch design, *J. Pharm. Sci.*, *85*(5), 491–495.
[57] Krishna, R. & Pandit, J. K. (1994). Transdermal delivery of propranolol, *Drug Dev. Ind. Pharm.*, *20*, 2459–2465.
[58] Jain, N. K. (2001). *Advances in Controlled and Novel Drug Delivery*, 1st Edit., CBS Publishers and Distributors, New Delhi, 108−110.
[59] Loyd, V. A. Jr., Popovich, N. G. & Ansel, H. C. (2005). *Pharmaceutical Dosage Forms and Drug Delivery Systems*, 8th Edit., Wolter Kluwer Publishers, New Delhi, 298−299.
[60] Chein, Y. W. (1992). Transdermal drug delivery and delivery system. *In: Novel Drug Delivery System*, Vol. *50*, Marcel Dekker, Inc., New York, 301−381.
[61] Willams, A. C. & Barry, B. W., (2004). Penetration enhancers, *Adv. Drug Del. Rev.*, *56*, 603−618.
[62] Pellet, M., Raghavan, S. L., Hadgraft, J. & Davis A. F. (2003). The application of supersaturated systems to percutaneous drug delivery. In: R. H. Guy and J. Hadgraft, (eds.), *Transdermal Drug Delivery*, Marcel Dekker, Inc., New York, 305−326.
[63] Brown, M. B. & Jones, S. A. (2000). Hyaluronic acid: A unique topical vehicle for localized drug delivery of drugs to the skin. *JEDV*, *19*, 308−318.
[64] Tsai, J. C., Guy, R. H., Thornfeldt, C. R., Gao, W. N., Feingold, K. R. & Elias, P. M. (1998). Metabolic approaches to enchance transdermal drug delivery, *J. Pharm. Sci.*, *85*, 643−648.
[65] Berner, B. & John, V. A. (1994). Pharmacokinetic characterization of transdermal delivery systems. *J. Clin. Pharmacokin.*, *26*(2), 121−134.
[66] Baker, W. & Heller, J. (1989). Material selection for transdermal delivery systems. In: *Transdermal Drug Delivery: Developmental Issues and Research Initiatives*, J. Hadgraft and R.H. Guys, (eds.), Marcel Dekker, Inc., New York, 293−311.
[67] Wiechers, J. (1992). Use of chemical penetration enhancers in Transdermal drug delivery-possibilities and difficulties. *Acta Pharm.*, *4*, 123.
[68] Yamamoto, T., Katakabe, K., Akiyoshi, K., Kan, K. & Asano, T. (1990). Topical application of glibenclamide lowers blood glucose levels in rats. *Diabetes Res. Clin. Pract.*, *8*, 19−22.
[69] Al- Khamis, K., Davis, S. S. & Hadgraft, J. (1986). Microviscosity and drug release from topical gel formulations. *Pharm. Res.*, *3*, 214−217.
[70] *Anon.*, (1980). Transdermal delivery systems-general drug release standards. *Pharmacopeial Forum*, *14*, 3860−3865.
[71] Mayorga, P., Puisieux, F. & Couarraze, G. (1996). Formulation study of a transdermal delivery system of primaquine. *Int. J. Pharm.*, *132*, 71-79.
[72] Deo, M. R., Sant, V. P., Parekh, S. R., Khopade, A. J. & Banakar, U. V. (1997). Proliposome-based transdermal delivery of levonorgestrel. *J. Biomat. Appl.*, *12*, 77−88.
[73] Yan-yu, X., Yun-mei, S., Zhi-Peng, C. & Qi-nerg, P. (2006). Preparation of silymarin proliposomes: A new way to increase oral bioavailability of silymarin in beagle dogs. *Int. Pharm.*, *319*, 162−168.
[74] Crawford, R. R. & Esmerian, O. K. (1997). Effect of plasticizers on some physical properties of cellulose acetate phthalate films. *J. Pharm. Sci.*, *60*, 312−314.

[75] Singh, J., Tripathi, K. T. & Sakia, T. R. (1993). Effect of penetration enhancers on the *in vitro* transport of ephedrine through rate skin and human epidermis from matrix based transdermal formulations. *Drug Dev. Ind. Pharm.*, *19*, 1623–1628.

[76] Wade, A. & Weller, P. J. (1994). *Handbook of Pharmaceutical Excipients*. Washington, DC: American Pharmaceutical Publ. Assoc., 362–366.

[77] Rhaghuram-Reddy, K., Muttalik, S. & Reddy, S. (2003). Once – daily sustained-release matrix tablets of nicorandil: Formulation and *in vitro* evaluation. *AAPS Pharm. Sci. Tech.*, *4*, 4.

[78] Shaila, L., Pandey, S. & Udupa, N. (2006). Design and evaluation of matrix type membrane controlled transdermal drug delivery system of nicotin suitable for use in smoking cessation. *Indian J. Pharm. Sci.*, *68*, 179–184.

[79] Aarti, N., Louk, A. R. M. P., Russsel, O. P. & Richard, H. G. (1995). Mechanism of oleic acid induced skin permeation enhancement *in vivo* in humans. *J. Controlled Rel.*, *37*, 299–306.

[80] Wade, A. & Weller, P. J. (1994). *Handbook of Pharmaceutical Excipients*. Washington, DC: American Pharmaceutical Publ. Assoc., 362–366.

[81] Lec, S. T., Yac, S. H., Kim, S. W. & Berner, B. (1991). One way membrane for transdermal drug delivery systems / system optimization. *Int. J. Pharm.*, *77*, 231–-237.

[82] Vyas, S. P. & Khar, R. K. (2002). *Targeted and Controlled Drug Delivery Novel carrier system*, 1st Edit., CBS Publishers and Distributors, New Delhi, 411–447.

*Chapter 8*

# INHALATION DELIVERY SYSTEMS

Inhalation is a very old method for drug delivery. In the 20th century, the method of inhalation became a mainstay for respiratory care, known as aerosol therapy. Use of inhaled epinephrine for relief of asthma was reported as early as 1929, in England. An early version of a dry powder inhaler (DPI) was the Aerohalor, used to administer penicillin dust to treat respiratory infections. In the 1950s, the Wright nebulizer was the precursor of the modern hand-held jet-venturi nebulizer. In 1956, the first metered-dose inhaler (MDI) was approved for clinical use, followed by the SpinHaler DPI for cromolyn sodium in 1971. The scientific basis for aerosol therapy developed relatively late, following the 1974 Sugarloaf Conference on the scientific basis of respiratory therapy. Early data on the drug-delivery efficiency of the common aerosol delivery devices (MDI, DPI, and nebulizer) showed lung deposition of approximately 10-15% of the total, nominal dose. Despite problems with low lung deposition with all of the early devices, evidence accumulated that supported the advantages of the inhalation route over other drug-administration routes. Inhaled drugs are localized to the target organ, which generally allows for a lower dose than is necessary with systemic delivery (oral or injection), and thus fewer and less severe adverse effects. The 3 types of aerosol device (MDI, DPI, and nebulizer) can be clinically equivalent. It may be necessary to increase the number of MDI puffs to achieve results equivalent to the larger nominal dose from a nebulizer. Design and lung-deposition improvement of MDIs, DPIs, and nebulizers are exemplified by the new hydrofluoroalkane-propelled MDI formulation of beclomethasone, the metered-dose liquid-spray Respimat, and the DPI system of the Spiros. Differences among aerosol delivery devices create challenges to patient use and caregiver instruction. Potential improvements in aerosol delivery include better standardization of function and patient use, greater reliability, and reduction of drug loss. An inhaler or puffer is a medical device used for delivering medication into the body via the lungs. It is mainly used in the treatment of asthma and chronic obstructive pulmonary disease. To reduce deposition in the mouth and throat, and to reduce the need for precise synchronization of the start of inhalation with actuation of the device, MDIs are sometimes used with a complementary spacer or holding chamber device. The most common type of inhaler is the pressurized metered-dose inhaler (MDI). In MDIs, medication is most commonly stored in solution in a pressurized canister that contains a propellant, although it may also be used as a suspension. The MDI canister is attached to a plastic, hand-operated actuator. On activation, the metered-dose inhaler releases a fixed dose of medication in aerosol form. The correct procedure for using an MDI is to first fully exhale, place the mouth-piece of the device into the mouth, and having just started to inhale at a moderate rate, depress the canister to release the medicine. The aerosolized medication is drawn into the lungs by continuing to inhale deeply before holding the breath for 10 seconds to allow the aerosol to settle onto the walls of the bronchial and other airways of the lung. Dry powder inhalers release a metered or device-measured dose of powdered medication that is inhaled through a DPI device. In 1968, Robert Wexler of Abbott Laboratories developed the Analgizer, a disposable inhaler that allowed the self-administration of methoxyflurane vapor in air for analgesia. The Analgizer consisted of a polyethylene cylinder 5 inches in length and 1 inch in diameter with a 1 inch long mouthpiece. The device contained a rolled wick of polypropylene felt which held 15 millimeters of methoxyflurane. Because of the simplicity of the Analgizer and the pharmacological characteristics of methoxyflurane, it was easy for patients to self-administer the drug and rapidly achieve a level of conscious analgesia which could be maintained and adjusted as necessary over a period of time lasting from a few minutes to several hours. The 15 millimeter supply of methoxyflurane would typically last for two to three hours, during which time the user would often be partly amnesic to the sense of pain, the device could be refilled if necessary. A simulated respiratory system was developed for the *in vitro* evaluation of two differently designed oral inhalation delivery systems. The deposition properties of a newly designed delivery system used for triamcinolone acetonide were compared to the more conventional, commercially available adapter utilized for an aerosol containing beclomethasone dipropionate. The simulated respiratory system was constructed so that the delivered dose of active ingredient could be classified into two fractions: the fraction that would be deposited in the oral cavity and throat and the fraction that would reach the desired site of activity in the respiratory tract. Based on this method, the newly designed system delivered more than 95% of the labeled dose to the desired site. The beclomethasone dipropionate aerosol system, which was observed to discharge the active ingredient with a greater intensity, delivered approximately 40% of the labeled dose. The particle-size distribution of the dose dispensed from the newly designed delivery system attached to the triamcinolone acetonide aerosol was determined using an impactor technique. No effort was made to correlate these results with an *in vivo* response. A medication inhalation therapy apparatus designed for the delivery of finely misted medicines to a patient in conjunction with the inhalation of 100% oxygen by the patient. The apparatus comprises a facemask equipped with an oxygen port for delivering oxygen, the facemask placed about or on the face of the patient. The facemask has a nebulizer mechanism disposed thereon, which is used to deliver medication in addition to oxygen. The nebulizer mechanism comprises a nebulizer port, which rotates 90 degrees in either direction to allow the present invention to be used in various angles. The nebulizer port is adapted to receive a nebulizer head, wherein are placed inhalable medications, which are nebulized. When the apparatus is used for medication inhalation therapy, the nebulizer head receives oxygen for nebulizing inhalable medications placed therein, the oxygen received from an oxygen port, which delivers oxygen to the facemask. When the apparatus is not being used for medication inhalation therapy, the nebulizer head is disconnected from the oxygen port. The apparatus may also be provided with a horizontal bag, which permits flow of oxygen to a patient, wherein the flow is unimpeded by objects placed upon the patient's chest. Effective therapies directed against asthma have advanced since 1956 when the first pressurized metered dose inhaler was developed. The target for these anti-asthma medications is the smooth muscle cells concentrated in the pulmonary airways. Delivery of drugs directly to the lungs is an effective and relatively safe means of both local and systemic treatment of a variety of medical conditions. Medication delivery may be inefficient if particles aggregate in the inhaler itself, if they are filtered by the throat and mouth, and if mucociliary clearance processes remove them. In order to keep dry powder aerosols from depositing in the oropharyngeal cavity and inhaler, most have a geometric diameter of 1-3 um and particle mass density of ~ $1 g/cm^3$. Unfortunately particles of this size also aggregate easily and are then phagocytosed by pulmonary macrophages resulting in a decrease in their therapeutic potential. Drug development for pulmonary use necessitates inhalation toxicology studies. Most of these studies require the use of relatively large drug amounts due to losses before and during exposure. A unique larger particle for inhalation use has been developed. These particles are porous with a large geometric diameter (~10 um) and low mass density ($<0.1 g/cm^3$). It has been hypothesized that these particles would be too large to be phagocytosed by macrophages; and would therefore remain in the lungs longer than conventional particles. An experimental model was developed using rats and guinea pigs to measure the efficacy of drug delivery using these particles. Both polymeric and nonpolymeric excepients were used to prepare these large particles with low mass density. Solvent evaporation and spray drying was used in both cases as described in detail in the following reference. The polymeric microsphere preparation was made of biodegradable poly(lactic acid-co-glycolic acid) (PLGA) polymer particles. The nonpolymeric large porous particles consisted of human serum albumin, lactose, dipalmitoyl phosphatidylcholine (DPPC), and albuterol sulfate.

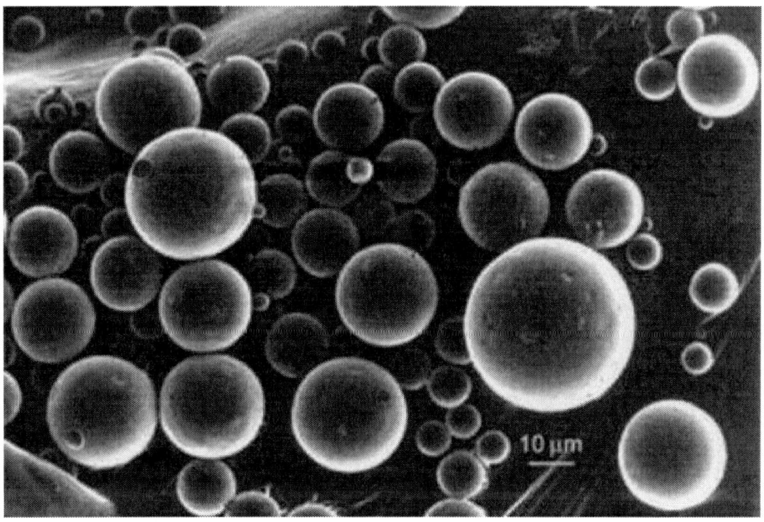


## PLGA Microspheres

Particles were also made of DPPC, albumin, lactose, and estradiol. The DPPC allows these nonpolymeric particles to be relatively water insoluble permitting a sustained release of hydrophilic or lipophilic drugs such as albuterol or estradiol over one to two days. The DPPC particles all had geometric diameters of ~10um and bulk (tapped) densities of approximately $0.08 g/cm^3$. Control non-porous particles made of lactose and albuterol had a smaller geometric diameter and larger bulk density (diameter=3u and bulk density of 0.45 g/cm$^3$). These particles were aerosolized with a Spinhaler dry powder inhaler and characterized *in vitro* with a cascade impactor and Aerosizer as detailed in the reference. About 50% of the large porous particles had a MMAD of <4.7 um while 16% of the small nonporous particles were <4.7 um or "fine particles." In addition, Spinhaler results showed that large porous particles were less aggregated than nonporous particles. The MMAD of large porous particles was ~2.15 um while the MMAD of nonporous particles was ~ 4.53 as measured with the Aerosizer. The smaller MMAD results in greater respiratory ability *in vitro*. The final test of the utility of large porous particles for both delivery and sustained release into the lungs required the development of a novel rodent exposure system. Rats were used to test the release over time of insulin, testosterone, and estradiol. Guinea pigs were used to assess the bronchial dilation effect of the albuterol. The animal is anesthetized and placed supine on a surgical board. The proximal trachea is exposed and a blunt cannula inserted and connected to a ventilator as shown in Figure 1.

The animal is still able to breathe through proximal and upper airways because the outer diameter of the cannula is 1 mm whereas the inner diameter of the trachea is 3-4 mm. (The 1mm diameter cannula has greater airway resistance than the 3-4 mm airway). A pipette with a rubber bulb is attached to the inhalation side of the ventilator tubing for the purpose of introducing the particles. The bulb is filled with a set amount of particles and a portion is introduced as the operator squeezes it while the respirator is set to inhale. The portion is delivered in one 30 to 60 sec insufflation by forced ventilation at 3 ml tidal volume and 100 strokes/min frequencies.

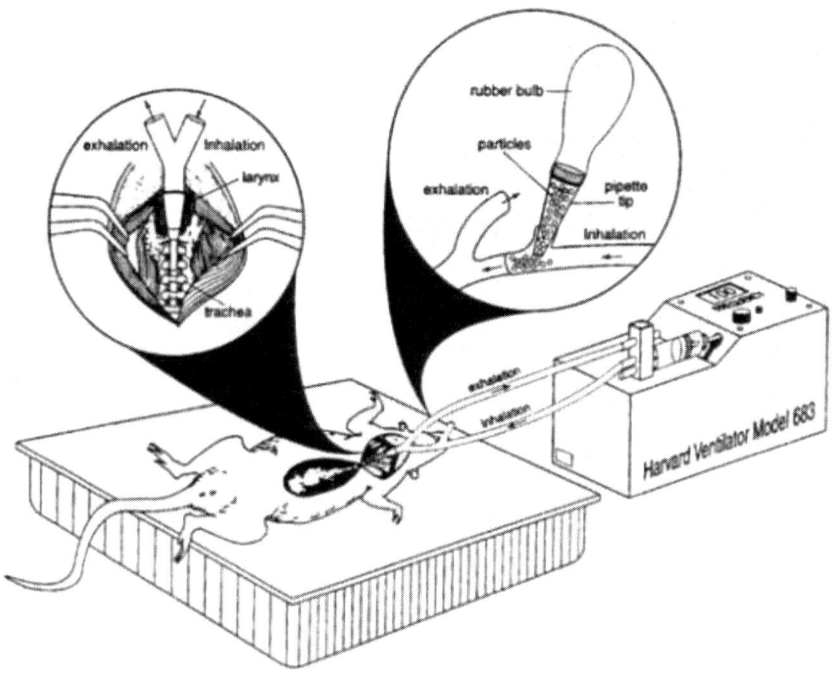


Figure 1. Experiment on bronchial dilation when administering abuterol.

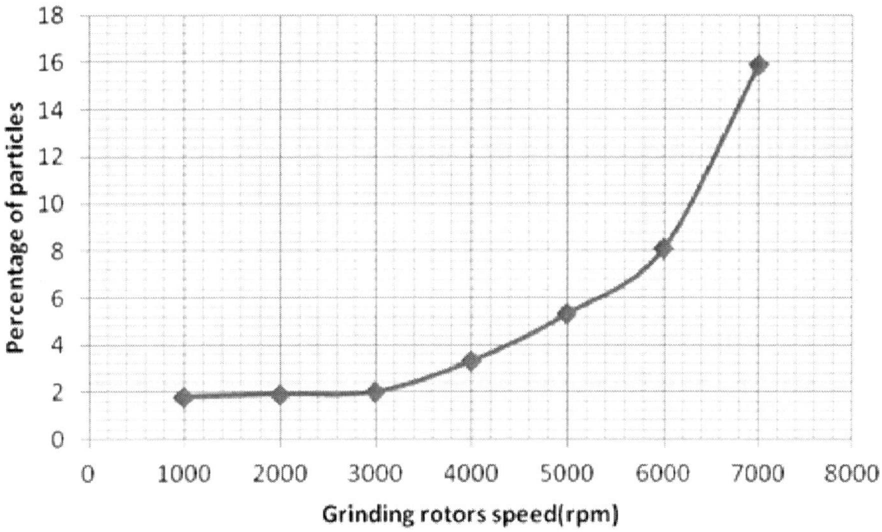


Figure 2. Grinding rotor air speed plots when using a classifier mill.

Several repetitions of the procedure deliver 3 to 5 mg of particles over a period of 10-15 min. The bronchoalveolar lavage fluid (BALF) was collected 10 min after the inhalation procedure and the number of particles in each lobe was counted as described in the reference. Figure 2 shows that most of the large porous particle mass was deposited in the lungs distal to the carina. Less than 45% was left in the tubing and trachea. However, only 20% of the mass of small nonporous particles was deposited in the respiratory tract. There were no significant changes in cell population or protein recovered from the BALF in guinea pigs treated with large porous particles with and without albuterol.

The bioavailability of the drugs, insulin, testosterone, and estradiol, administered to the lungs with large porous particles was compared to the bioavailability when administered subcutaneously. Radioimmunoassay was used to quantitatively measure the amount of insulin, estradiol, and testosterone. The results had 87.5% more estradiol, 177% more insulin, and 86% more testosterone were available from inhaled large porous particles compared to subcutaneous administration of these drugs. In contrast the small nonporous particles yielded less bioavailable insulin (12%) and estradiol (18%). Blood concentrations of these drugs remained elevated significantly longer when they were administered on large porous particles compared to administration on nonporous particles. There was also a sustained inhibition of carbachol-induced bronchoconstriction when guinea pigs inhaled large porous particles with albuterol for 15 min. The authors concluded that large porous particles reach the deep lung more efficiently than do more conventional small nonporous particles in this experimental rodent model. In addition, large porous particles have a longer residence time in the lungs than do small nonporous particles. The large porous particles have relatively low mass density. This causes better penetration of the large porous particle because the aerodynamic diameter of these particles is similar to that of nonporous particles. Greater aerosolization efficiency allows less chance of loss in large airways and tubing. The insufflation technique for aerosol delivery described is concluded to be useful for dry powders because it allows a small amount of aerosol to be delivered in a short period of time. Many of the drugs used to treat lung diseases are administered to the patient through the lungs. The advantages of this route include the ability to use lower doses, fewer side-effects and the early onset of drug activity. This method of drug administration is increasingly being used for systemic drug applications, particularly when injection is the only alternative. The traditional dosage form for inhalation was the metered dose inhaler (MDI). In these devices, the drug was dispersed by chlorofluorocarbons (CFCs) followed by inhalation.

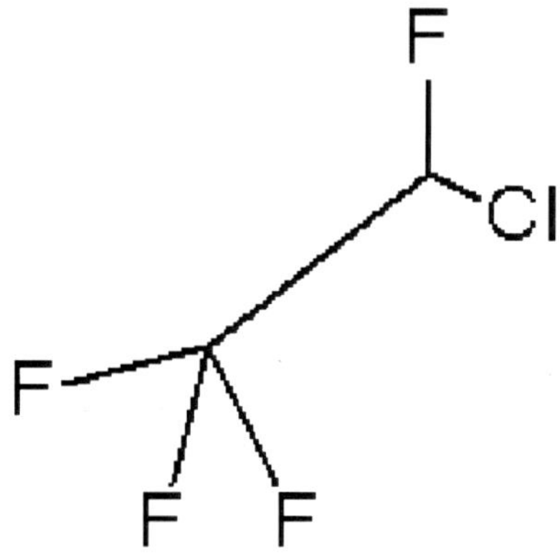


## CFC Molecules

Because of environmental implications, however, CFCs were banned, leading to MDIs partly being superseded by dry powder inhalers (DPIs).

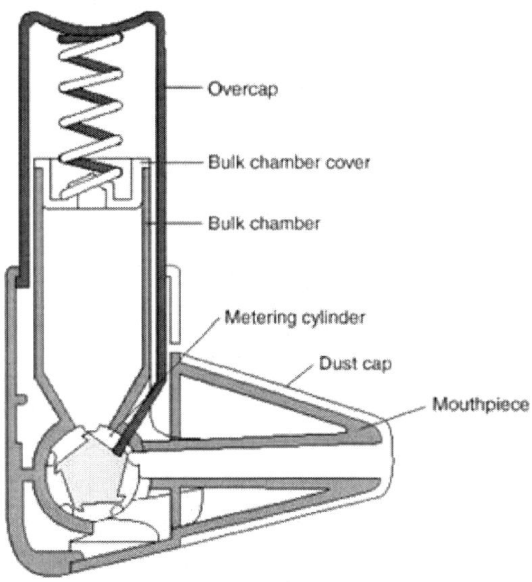


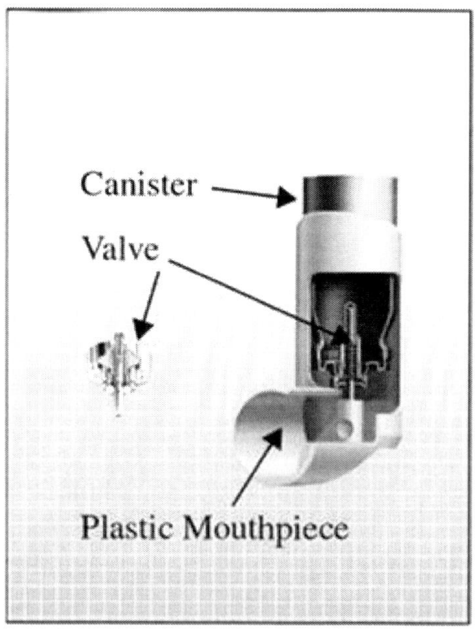


Source: *Clin. Drug Invest.*- Adis International Limited.

Dry Powder Inhalers.

DPIs contain a powder consisting of a drug and a carrier. In some systems the powder is packed in capsules or blisters that contain one dose, and are therefore referred to as single-dose systems. Other devices contain a powder reservoir, known as a multi-dose system. The most popular carrier used is lactose. Toxicology data of lactose after administration to the lungs were favorable. Moreover, lactose is readily available and well established, since it has been used in DPIs for more than 20 years. Drug delivery depends on the breathing behavior of the patient. Regardless of a patient's breathing action, it is essential that the dose is released in exactly the same way every time. Consistent delivery at different airflows is therefore essential for a DPI and is assisted by the presence of a carrier in the formulation. Generally, DPI powders have the ability to flow well. The flow ability has to be sufficient both for the correct DPI dosing and manufacture. These three requirements dictate the formulation of the powder to be inhaled, of which lactose is a major constituent. Therefore, lactose for inhalation has to meet some basic requirements, as discussed in this article. Inhalation when using a DPI will reach the lungs by the following route: 1) The drug is discharged from the capsule or blister (single-dose system) or a dose is taken from the reservoir (multi-dose system). This action is done mechanically, by pushing a button on the DPI for example. 2) The user inhales during discharge. The action of breathing provides the airstream for the further transport of the powder. 3) The drug separates from the lactose, and the drug particles enter the respiratory system. 4) Typically, the lactose particles are too large to enter the respiratory system. They will remain in the throat. Only when lactose particles are as small as the drug particles will they enter the lungs. 5) The location in the respiratory tract where the drug is delivered depends upon its particle size. Drug particles of 2-5 mm are delivered to the lungs. Larger particles do not fully penetrate the respiratory tract, whereas particles smaller than 2 mm are exhaled. Most drug particles between 2-5 mm tend to agglomerate when they are not formulated with a carrier and will not, therefore, reach the lungs. To prevent this agglomeration, the drug is formulated with lactose (or another carrier) to form an adhesive mixture. 6) This mixture uses the following principle: agglomeration of micronized drugs is caused by the hydrophobic character and large surface area of their particles. The attraction to each other results in an agglomerate; however, blending with a carrier prevents this agglomeration. Thus, micronized drug particles are attached to the surface of a lactose carrier. 7) Generally, such blends are formed when the lactose particles are at least five times larger than the drug particles. Additionally, there should be at least five times the amount of lactose compared with the drug. 8) Advantages of lactose particles make suitable carriers because of their regular shape and smooth surface, and it is these factors of favorable shape, size, and surface properties that provide its superior flow ability. Every drug has its own characteristics regarding a tendency to agglomerate and the ability to form an adhesive mixture with lactose. Moreover, every type of inhalation device has its own characteristics regarding both production and use by the patient. 9) Therefore, for every single combination of drug and DPI device, and specific lactose, often with a narrow particle size distribution, is required. Thus, for every new drug/DPI device combination, a new form of lactose has to be developed. Changes in the performance of the DPI can be achieved either through changes in the design of the device through changes in the powder formulation, or as described extensively, the forces governing the particle–particle interactions in the agglomerates and the forces playing a role in the de-agglomeration process. Recent developments regarding the powder formulation aim at a reduction of the adhesive and cohesive forces between the particles to increase the FPF. Supercritical fluid technology is applied to improve the surface properties of the drug substance. The large porous particles have reduced inter-particulate forces because of their low density, the irregular surface structure and/or reduced surface free energy. Moreover, these particles are claimed to have improved aerodynamic behavior in the airways, whereas phagocytosis of the deposited particles in the alveoli is reduced. In another approach, smaller porous particles (3–5 mm) have been used to improve de-agglomeration and lung deposition.

Changes in the latest DPI device technologies provide new developments that aims at increasing the de-agglomeration forces generated during inhalation. It is well known that the more efficient the force is, higher the FPF will be. An important classification parameter for these new device developments is whether or not the powder de-agglomeration is power assisted (active devices) or must it depend on the kinetic energy of the inhalation flow generated by the patient (passive devices). Regarding the passive devices, recently two DPI devices were introduced that apply impaction forces for the aerosol generation.

## REFERENCES


[1] Bartow, R. A. & Brogden, R. N. (1998). Formoterol. An update of its pharmacological properties and therapeutic efficacy in the management of asthma. *Drugs*, *55*, 303–322.

[2] Beasley, R., Crane, J. & Burgess, C. D. (2000). Long-acting h2-agonist drugs. *Lung Biol. Health and Dis.*, *139*, 119–154.

[3] Bennett, J. A. & Tattersfield, A. E. (1997). Time course and relative dose potency of systematic effects from salmeterol and salbutamol in healthy subjects. *Thorax*, *52*, 458–464.

[4] Bremner, P., Woodman, K., Burgess, C., Crane, J., Purdie, G., Pearce, N. & Beasley, R. (1993). A comparison of the cardiovascular and metabolic effects of formoterol, salbutamol and fenoterol. *Eur. Respir. J.*, *6*, 204– 210.

[5] Brodde, O. E., Schuler, S., Kretsch, R., et al. (1986). Regional distribution of beta-adrenoceptors in the human heart: Coexistence of functional beta1- and beta2-adrenoceptors in both atria and ventricles in severe congestive cardiomyopathy. *J. Cardiovascular Pharmacol.*, *8*, 1235–1245.

[6] Brown, J. E., McLeod, A. A. & Shand, D. L. (1986). In support of cardiac chronotropic beta2 adrenoceptors. *Amer. J. Cardiol.*, *57*, 11 – 6F.

[7] Carlsson, L., Abrahamsson, C., Andersson, B., Duker, G. & Schiller- Lindhardt, G. (1993). Proarrhythmic effects of the class III antiarrhythmic agent almokalant: Importance of infusion rate, QT dispersion and early after depolarizations. *Cardiovascular Res.*, *27*, 2186–2193.

[8] Cazzola, M., Imperatore, F., Salzillo, A., Di Perna, F., Calderaro, F., Imperatore, A. & Matera, M. G. (1998). Cardiac effects of formoterol in patients suffering from COPD with preexisting cardiac arrhytmias and hypoxemia. *Chest*, *114*, 411–415.

[9] Cuddihy, R. G., Brownstein, D. G., Raabe, O. G. & Kanapilly, G. M. (1973). Respiratory tract deposition of inhaled polydispersed aerosols in beagle dogs. *J. Aerosol Sci.*, 5, 35–45.

[10] Clapham, J. C. & Hamilton, T. C. (1992). Effects of BRL38227, salbutamol & aminophylline alone and in combination, on plasma potassium and on the heart. *Drug Dev. Res.*, *26*, 157–172.

[11] Ganderton, D. (1999). Chairman's summary. *J. Aerosol Medicine*, *12*(1), 41–42.

[12] Guhan, A. R., Cooper, S., Oborne, J., Lewis, S., Bennett, J. & Tattersfield, A. E. (2000). Systemic effects of formoterol and salmeterol: A dose– response comparison in healthy subjects. *Thorax*, *55*, 650–656.

[13] Hochrainer, D. & Hölz, H. (2001). Comparison of cloud velocities delivered from RespimatR soft mist inhaler or MDIs. *J. Aerosol Medicine*, *14*(3), 1-5.

[14] ICH S7a (2001). Guidance for industry S7A safety pharmacology studies for human pharmaceuticals. FDA. Lecaillon, J.B., Kaiser, G. and Palmisano, M. (1999). Pharmacokinetics and tolerability of formoterol in health volunteers after a single high dose of Foradil dry powder inhalation via aerolizer. *Eur. J. Clin. Pharmacol.*, *55*, 131–139.

[15] Matsunaga, T., Mitsui, T., Harada, T., Inokuma, M., Murano, H. & Shibutani, Y. (1997). QT corrected for heart rate and relation between QT and RR intervals in beagle dogs. *J. Pharmacol. & Toxicol. Methods*, *38*, 201–209.

[16] Moore, R. H., Khan, A. & Dickey, B. F. (1998). Long acting inhaled b2-agonist in asthma therapy. *Chest*, *113*, 1095–1108.

[17] Newmann, S. P. (1999). Use of gamma scintigraphy to evaluate the performance of new inhalers. *J. Aerosol Medicine*, *12*(1), 25–31.

[18] Redington, A. E. & Rees, P. J. (1998). Long-acting inhaled b2-agonists in the management of asthma: Recent advances and current recommendations. *Intl J. Clin. Pract.*, *57*, 482–486.

[19] Sarma, J. S. M., Sarma, R. J., Bilitsch, M. & Katz, D. (1984). An exponential formula for heart rate dependence of QT interval during exercise and cardiac pacing in humans: Reevaluation of Bazett's formula. *Amer. J. Cardiol.*, *54*, 103–108.

[20] Strauss, M. H., Reeves, R. A., Smith, D. L., et al. (1986). The role of cardiac beta-1 receptors in the thermodynamic response to a beta-2 agonist. *Clin. Pharmacol. Ther.*, *40*, 108–115.

[21] Tattersfield, A. E. (1993). Clinical pharmacology of long-acting-receptor agonists. *Life Sci.*, *52*, 2161–2169.

[22] Van Noord, J. A. (2000). Delivery of fenoterol via RespimatR. *Respiration*, *67*, 672–678.

[23] Vaugham Williams, E. M. (1985). Cardiac electrophysiologic effects of selective adrenoceptor stimulation and their possible roles in arrhythmias. *J. Cardiovascular Pharmacol.*, *7*(suppl. 5), 61– 64.

[24] Vincken, W. J., Aumann, M., Radermecker, H., Slabbynck, I., Stappaerts, M., Demedts, M. & Moonen, D. (1997). Comparison of pharmacodynamics, efficacy and safety of fenoterol inhaled from a novel pocket-sized inhaler and from a metered dose inhaler (MDI). *Eur. Respir. J.*, *10*(Suppl. 25), 239.

[25] Wolff, R. K., Kanapilly, G. M., Gray, R. H. & McClellan, R. O. (1984). Deposition and retention of inhaled aggregate 67Ga2O3 particles in beagle dogs, Fischer-344 rats & CD-1 mice. *Amer. Indust. Hygiene Assoc. J.*, *45*, 377–381.

[26] Zierenberg, B. (1999). Optimizing the *in vitro* performance of Respimat. *J. Aerosol Medicine*, *12*, 19–24.

*Chapter 9*

# Pulmonary Delivery Systems

Pulmonary route has been used to treat various respiratory diseases for centuries. Ancient inhalation therapies included the use of leaves from plants, vapors from aromatic plants, balsams, and myrrh. However, around the turn of the 19th century, with the invention of liquid nebulizers, these early treatments developed into legitimate pharmaceutical therapies. In the 1920s, adrenaline was introduced as a nebulizer solution. In 1925, nebulizer porcine insulin was used in experimental studies in diabetes. In 1945, pulmonary delivery of the recently discovered penicillin was investigated. Steroids had been introduced in the mid-1950s for the treatment of asthma, and nebulizers were enjoying widespread use. In 1956, the pressured metered dose inhaler (pMDI) was introduced. Over the past five decades, by the advances in molecule design and drug discovery the pMDI has risen to become the main stay of asthma treatment.

Over the last decade certain drugs have been sold in compositions suitable for forming drug dispersion for pulmonary delivery to treat various conditions in humans. Such pulmonary drug delivery compositions were designed to be delivered to the patient by inhalation so that the active drug within the dispersion can reach the lungs. It has been found that certain drugs given by pulmonary route are readily absorbed through the alveolar region directly into the blood circulation. The pulmonary route possesses many advantages over the other routes of administration for the treatment of specific disease states. Particularly because large protein molecules which degrade in the gastrointestinal conditions and are eliminated by the first pass metabolism in the liver can be delivered via the pulmonary route if deposited in the respiratory zone of the lungs. Devices that can be used to deliver a drug by pulmonary route area are based on one of three platforms: pressurized metered dose inhaler, nebulizer, and dry powder. In the treatment of disease, aerosol administration represents a valuable means by which a therapeutic agent may be delivered.

## Ideal Characteristics of Therapeutic Aerosol

Therapeutic aerosol is a safe and efficacious drug that contains minimal quantities of inert excipients. Aerosol drug has the ability to monodisperse since it has a small particle size with low velocity after generation; high concentration and with a high rate of generation, and highly reproducible characteristics for low bioburden (solids) or sterile (liquids).

## ADVANTAGES OF PULMONANRY DRUG DELIVERY

Inhaled drug delivery puts the drug where it is needed. It requires low or fractional oral dose, i.e., drug content of one 4 mg tablet of salbutamol equals to 40 doses of meter doses. Pulmonary drug delivery has a very negligible side effect since the rest of body is not exposed to drug.

The onset action is also very quick as a pulmonary drug delivery. Another advantage is that the degradation of the drug by the liver is avoided in a pulmonary drug delivery. Asthma and diabetes requires long term treatment if it is given by pulmonary drug delivery which allows the patient's safety to be at maximum because the rest of body is not exposed to the drug.

## CHALLENGES IN PULMONARY DRUG DELIVERY

### Low Efficiency of Inhalation System

Efficiency of presently available inhalation systems has been generally too low which is an important challenge regarding pulmonary drug delivery. Optimum aerosol particle size is very important for deep lung delivery.

Optimum particle size for deep lung deposition is 1-5 mm. Aerosol system should have to produce optimum size particles because if too small, the particles will be exhaled. If the particles are too large, then this can affect the oropharynx and larynx.

### Less Drug Mass Per Puff

To get adequate effect with the pulmonary drug delivery, practical delivery of many drugs will require milligram doses. Yet, in most existing systems, the total amount of drug per puff delivered to the lower respiratory tract is, in fact, too low; less than 1000 mcg.

### Poor Formulation Stability for Drug

Most traditional small molecule asthma drugs are crystalline. In the case of corticosteroids, it is relatively moisture resistant in the dry macromolecules. For corticosteroids, they often become unstable in the liquid state, amorphous, and highly moisture sensitive in the dry state.

### Improper Dosing Reproducibility

There are various factors for poor dosing reproducibility, as this possibility could worsen the patient's disease if there were problems with the device, or result of an instability in the drug formulation. To get maximum dose reproducibility for the patient, then patient education must play an important role.

## Recent Advances in Pulmonary Drug Delivery

Recent advances in the technology of pulmonary drug delivery involved new methods or key concepts in delivery, formulation, broader applications, and the goal to develop better devices. Partly through adaptation of these key concepts, great strides have been made in the field of pulmonary drug delivery. Over the past 15 years, inhalation systems have been developed that make use of these new technological concepts.

### Improvements Regarding the Fine Particle Fraction (FPF)

By using an inhalation system that produces particles of 1-5 mm size and fine particle fraction will produce good deposition in the lungs. A high batch-to-batch reproducibility of the FPF can reduce variability in lung deposition. There are advantages in using the monodisperse inhalation system to target the deposition in the lungs.

### Improvements that Affect Particle Velocity and Reduce Coordination Problems

To avoid oropharyngeal deposition, systems that can produce the aerosol at a reduced velocity is preferred. Systems that not only release the aerosol during the first 0.5 s of the inhalation, but also generate the aerosol over a period longer than 1 s, will reduce the coordination problems for the patient. The coordination problems for a patient is reduced by the benefits of breath actuated systems.

### Improvements Reduce the Cost and Increase the Patient Compliance

To further enhance this type of aerosol system for these selected systems, it is important that they are simple to use and robust, nonfragile, small, and easy-to-carry; and do not require electricity or pressurized air.

## Recent Technologies Use in Pulmonary Drug Delivery

### Particle Engineering for Pulmonary Drug Delivery

Recent advances in inhalation therapy have sparked considerable biomedical interest in the development of novel particle technologies for respiratory drug formulation. Introduction of new potent medicines in various therapeutic areas such as asthma, chronic obstructive pulmonary disease (COPD), and various infectious diseases has necessitated an accurate and consistent dosing with inhalation devices. There are many emerging inhalation products with new absorption mechanisms and/or rapid onset of action for systemic therapies. Controlled and sustained release with composite particles is another application that can be used for both local and systemic drug deliveries.

### Agglomerated Vesicle Technology for Pulmonary Drug Delivery

These particles are multimicron-sized nanoparticles that are composed of chemically linked agglomerates. The links between the nanoparticles can be either permanent (e.g., carbonyl) or cleavable (e.g., disulfide or ester). Complex agglomerate structures can be achieved by scheduling the application of linker agents. The release rate of drugs from the assembly can be modulated by controlling the extent of cleavage in the links. One envisions the structure of the agglomerate during cleavage being controlled by the location of the permanent and cleavable links. In agglomerate experiments the release of ciprofloxacin from these agglomerates *in vitro* are presented. Recent developments regarding the powder formulation aim at a reduction of the adhesive and cohesive forces between the particles to increase the FPF.

## Recent Advances in Pulmonary Drug Delivery Devices

The types of inhalation devices presently available and used in for drug delivery are the metered dose inhalers, the dry powder inhalers, or nebulizer.

### Inhalation Drug Delivery System By Metered Dose Inhalers

A metered-dose inhaler (MDI) is a complex system designed to provide a fine mist of medicament, generally with an aerodynamic particle size of less than 5 microns, for inhalation directly to the airways for the treatment of respiratory diseases such as asthma and COPD.

## Advances in MDI Technology and Use in Enantiomer Preparations of Inhaled Drugs

There has been much interest in the effective differences of Enantiomer with many other kinds of medications. Beta agonist adrenergic bronchodilators had also received much attention regarding its effectiveness. Recently, levo salbutamol active enantiomer which is available on the present market is free from termers and palpitation that are often seen in salbutamol. Similarly, the (R)-enantiomer of albuterol is mainly responsible for bronchodilator while the (S)-enantiomer may stimulate airway reactivity. Data suggests, however, that after aerosol delivery, the systemic absorption for (R)-albuterol is faster than for (S)-albuterol and, conversely, the lung retention of (S)-albuterol is longer, which may be detrimental.

## GENERIC PROLIFERATION IN DEVICES AND MEDICATIONS

Nowadays, new MDI and nebulizer brands are being introduced regularly in the pharmaceutical market. Even for those who watch this field, it is not unusual to hear a new, unfamiliar brand named regularly.

One trend has been the move to generic MDIs and for over the counter availability. These are introduced in the literature by comparing them with well-known older devices. Often, documentation that generic brands or new devices are comparable to older ones is difficult to come by, so comparisons showing pharmacokinetic equivalency are useful.

## NEW TECHNOLOGIES TO IMPROVE PATIENT INHALATION COORDINATION WITH MDI

Spacers are used to improve patient coordination with MDI. Both adults and children often have difficulty coordinating the inhalation effort with the timing of the aerosol puff. Evidence indicates considerable intra and intersubject variability for the inhalation technique.

## FLOW GATE VALVE TECHNOLOGY IN SPACERS

Certain companies manufacture MDIs with static free spacers that allow the valve mechanism to increase the drug dose reaching the lungs. Valves can open during inhalation and then close during exhalation to ensure that the residual dose is retained in the spacer for subsequent inhalation.

## THE AUTOHALER MODIFIED FORM OF pMDI

The Autohaler™ is the first breath actuated or activated pressurized metered dose inhaler. Autoihaler™ solved the key problem of the pressurized metered dose inhaler (pMDI) viz. coordination of actuation with inhalation and does not rely on the patient's inspiratory effort to aerosolize the dose of medication unlike dry powder inhalers. The Autohaler ™ is a modified form of pressurized metered dose inhaler.

### Inhalation Drug Delivery Device by Dry Powder Inhalers

Today there are essentially two types of DPIs, those that use drug-filled discrete individual doses, e.g., either a gelatin capsule or a foil–foil blister, and those that use a reservoir of drug that meters out doses when required. Both are now widely available around the globe and are gaining broad acceptance. Unit dose devices for single-dose powder inhalers are devices in which a powder containing capsule is placed in a holder. The capsule is opened within the device and the powder is inhaled. The capsule residue must be discarded after use and a new capsule inserted for the next dose. For multidose devices, there is a circular disc that contains either four or eight powder doses on a single disc. This typically would be treatment for one to two days. The doses are maintained in separate aluminum blister reservoirs until just before inspiration. This device is a true multidose device, to include 60 doses in a foil–foil aluminum strip that is opened only at the point just prior to patient inspiration.

## New Developments in Dry Powder Inhalation Technology

The changes in performance for the DPI can be achieved either through modifying the design of the device or through changing the powder formulation. Or as described extensively, to control the forces which govern the particle–particle interactions in the agglomerates and the forces that play a role in the de-agglomeration process. Recent developments regarding the powder formulation aim at a reduction in the adhesive and cohesive forces between the particles to increase the FPF. Supercritical fluid technology is applied to improve the surface properties of the drug substance. Thus, by using large porous particles to reduce the inter-particulate forces because of their low density, the irregular surface structure and/or reduced surface free energy. Moreover, these particles claim to improve aerodynamic behavior in the airways, whereas the phagocytosis deposited particles in the alveoli have shown the opposite. In another approach, smaller porous particles (3–5 mm) have been used to improve de-agglomeration and lung deposition.

Current device technologies for these new developments really aim to increase the de-agglomeration forces generated during the inhalation. It is well known that the more efficient the force is, higher the FPF will be. A major classification parameter in the new device developments is whether or not the powder de-agglomeration is power assisted (active devices) or depends on the kinetic energy of the inhalation flow generated by the patient (passive devices). Regarding the passive devices, recently two DPI devices were introduced that apply impact forces for the generation of the aerosol.

### Airclassifier Technology in Devices

An airclassifier is another important technology used in recent devices for pulmonary drug delivery. The inhaler contains a classifier (cyclone) chamber in which high inertial forces are applied onto the rotating particles. Moreover, the action of these forces on the larger agglomerates is sustained because they remain in the classifier for a certain period of time, which can be controlled by the classifier design and by the choice of carrier-size fraction. Airclassifier technology system contains a cyclone chamber for particle de-agglomeration. Modified form of the airclassifier technology is the multiple airclassifier technology.

## Multiple Airclassifier Technology

The technology used for a multiple airclassifier is that the chambers are placed in a parallel arrangement, which further increases the dose that can be aerosolized. Another interesting point in this development is the invention of the disposable DPI. The concept of a disposable inhaler is interesting because it reduces the chance of microbial contamination.

## Inhalation Drug Delivery Devices by Nebulizer

Mainly there are two general types of nebulizer systems, the ultrasonic and the air jet. In ultrasonic nebulizers, ultrasound waves are formed in an ultrasonic nebulizer chamber by a ceramic piezoelectric crystal that vibrates when electrically excited. These vibrations set up high energy waves in the solution, within the device chamber, of a precise frequency that generates an aerosol cloud at the solution surface. The aerosol produced by an air jet nebulizer is generated when compressed air is forced through an orifice. An area of low pressure is then formed where the air jet exists. A liquid may be withdrawn from a perpendicular nozzle (the Bernoulli Effect) to mix with the air jet to form droplets. A baffle (or baffles) within the nebulizer is often used to facilitate the formation of the aerosol cloud. Carrier air can be used to generate the "air jet." Alternatively, compressors may be used to generate the air stream. Nebulizers used today for drug delivery to the respiratory tract and are particularly useful for the treatment of hospitalized or nonambulatory patients.

Recent developments in liquid aerosol technology combine the advantages of MDIs and nebulizers, called metered-dose liquid inhalers. As we know that the CFC propellants were replaced by HFAs in MDIs and the classical nebulizer area was extended with the adaptive aerosol delivery (AAD) technology in the 90s during the past century. These systems try to combine the advantages that MDIs and nebulizers offer, and to get rid of the many disadvantages they suffer from.

The major advantage that all of these systems aim for is to reduce the velocity of the aerosol. This feature is of paramount importance because it will reduce oropharyngeal deposition and increase the lung deposition. Liquid inhalers that apply the concept of a low velocity aerosol are often referred to as 'soft mist inhalers.' Further developments in the field of wet nebulization aim at the generation of monodisperse aerosols, in the absence of propellants in the formulation by applying aqueous drug formulations. Where there is a reduction in the residual volume after nebulization and an improved portability compared with nebulizers.

Aerosol usage in new technologies are classified into four systems that force the liquid through a nozzle - Respimat, AERx™, MedSpray Systems that use a vibrating mesh or perforated plate - Aerodose, Touchspray™ technology. In the use of electrohydrodynamic breakup to generate the aerosol, Mystic™ (Battelle) Systems apply condensation to generate the aerosol.

# Recent Advances of Formulation in Pulmonary Drug Delivery

Effective inhalable medications are produced by drug formulation. Formulation stability is another challenge for producing pulmonary drug delivery. Formulation stability is responsible for keeping the drug pharmacologically active; it must be efficiently delivered into the lungs, to the appropriate site of action, and to remain in the lungs until the desired pharmacological effect occurs. Depending upon disease condition, the effective formulation release drug, such as insulin for diabetes, must be deposited in the lung periphery to ensure maximum systemic bioavailability. Thus, a formulation that is retained in the lungs for the desired length of time and avoids the clearance mechanisms of the lung may be necessary. Research into dry powder formulations has been an area of growth in recent years and will be our focus in this section. Various techniques are used in dry powders formulation for inhalation that involves either, micronization via jet milling, precipitation, or spray-drying using various excipients, such as lipids and polymers, or carrier systems like lactose.

## Lactose Carrier Systems

Recent advances in inhalation therapy have sparked considerable biomedical interest in the development of novel particle technologies for respiratory drug formulation .The cohesive powders with poor flow arises if the surface electric forces associated with the particles exceed the gravitational force acting upon them. To overcome this problem, the drug is blended with a coarse carrier system (30–100 μm), such as lactose. At present, marketed dry powder inhalers contain either the drug alone or mixed with a bulk carrier, usually lactose (α-lactose monohydrate). Lactose has an established safety profile and improves the flow properties of the formulation necessary for reproducible filling and promoting dosing accuracy.

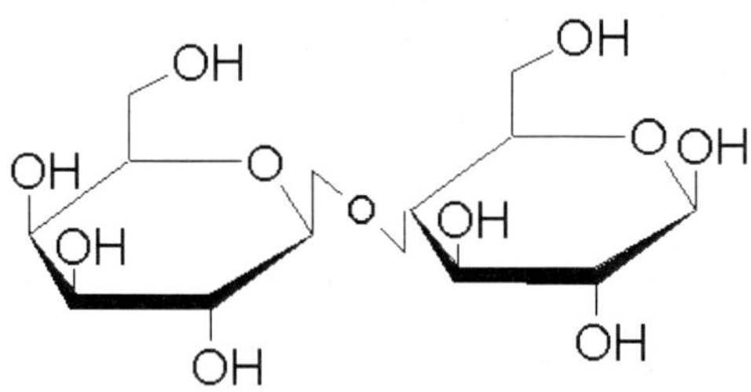


Lactose

## Liposomes

Liposomes as a pulmonary drug delivery vehicle has been studied for years and used as a means of delivering phospholipids to the alveolar surface for treatment of neonatal respiratory distress syndrome. More recently, liposomes have been investigated as a vehicle for sustained-release therapy in the treatment of lung disease.

## Large Porous Particles

Pulmospheres are a new type of aerosol formulation that has large porous hollow particles. The pulmospheres have low particle densities, excellent dispersibility and can be used in both MDI and DPI delivery systems. These particles can be prepared using polymeric or nonpolymeric excipients, by solvent evaporation and spray-drying techniques. Pulmospheres are made of phosphatidylcholine, the primary component of human lung surfactant. The large size of pulmospheres allows them to remain in the alveolar region longer than their nonporous counterparts by avoiding phagocytic clearance. After intratracheal administration into rats, only 8% and 12.5% of macrophages contain pulmospheres particles immediately and 48 h after inhalation, respectively; compared with 30% and 39% of macrophages containing nonporous particles during a similar time interval.

## Biodegradable Polymers

Apart from Liposomes, biodegradable polymer microspheres are currently being studied as sustained release pulmonary drug carriers. Polymers such as polylacticacid used in medical applications such as sutures orthopedic implants and medical dressings, and poly glycolic acid have been investigated. Also, polymers can be used mainly as a therapy or method of delivering therapeutic agents to the alveolar surface for the treatment of systemic diseases. Sustained release from a therapeutic aerosol can prolong the residence of an administered drug in the airways or alveolar region, minimize the risk of adverse effects by decreasing its systemic absorption rate, and increase patient compliance by reducing dosing frequency. A sustained-release formulation must avoid the clearance mechanisms of the lung, the mucociliary escalator of the conducting airways and macrophages in the alveolar region. Only a limited amount of research has been published in this area; although the sustained-release profiles achieved with corticosteroids appear promising. However, the toxicity of this type of formulation has not yet been established for lung delivery.

## Advances in Propellants Used in Pulmonary Drug Delivery Devices

Recently, HFA propellants are the new alternative for CFC propellants in pulmonary drug delivery devices. Generally the higher vapor pressures of the HFAs (particularly 134a) have the potential to generate aerosols of higher quality than the "old" CFC formulations. However, except in a few notable instances, potential product improvements have been sacrificed for development costs and time as per market need.

## NEW APPLICATIONS OF PULMONARY DRUG DELIVERY

Apart from asthma and COPD, recently pulmonary drug delivery is used for the following conditions: insulin by aerosol treatment of migraine, nicotine aerosol for Smoking, cessation aerosols for angina, aerosol vaccination, alpha 1 antitrypsin aerosols in transplantation pulmonary arterial hypertension, acute lung injury surfactant, aerosol gene therapy via aerosol in cancer chemotherapy, pentamidine aerosol amphotericin B, gentamycin aerosol, ribavirin aerosol zanimivir R/C with revolizer for swine flu. Aerosols are used in clinical investigations of disease inhaled drug delivery for tuberculosis therapy pulmonary delivery for lower molecular weight heparin. Aerosol therapy also includes the controlled delivery for drugs to the lungs for pulmonary delivery of drugs for bone disorders, and pulmonary delivery of opioids as pain therapeutics.

## REFERENCES

[1] Newhouse, M. T. (2000). *Encyclopedia of Pharmaceutical Technology*, 2nd Edit., Dekker, New York Informa Healthcare USA, *19*, 1279-1285.

[2] Daniher, D. I. & Zhu, J. (2008). Review on "dry powder platform for pulmonary drug delivery," www.sciencedirect.com, *Particuology*, 225–238.

[3] www.cipladoc.com.

[4] Frijlink, H. W. & de Boer, A. H. (2005). Trends in the technology-driven development of new inhalation devices, *Drug Discovery Today Technology*, Vol. *2*, No.1.

[5] Chow, A. H. L., Tong, H. H. Y., Chattopadhyay, P. & Shekunov, B. Y. (2007). Particle engineering for pulmonary drug delivery, *Pharm. Res.*, *24*(3), 411−433.

[6] Edwards, D. A., Ben-Jebria, A. & Lange, R. (1998). Recent advances in pulmonary drug delivery using large, porous inhaled particles, *J. Appl. Physiol.*, *85*(2), 379−385.

[7] Bhavane, R., Karathanasis, E. & Annapragada, A. V. (2003). Agglomerated vesicle technology: A new class of particles for controlled and modulated pulmonary drug delivery, *J. Controlled Rel.*, 15– 28.

[8] Atkins, P. J. & Crowder, T. M., Atkins, P. J. & Crowder, T. M. *The Design and Development of Inhalation Drug Delivery Systems*, Modern Pharmaceutics by Marcel Dekker, 1−31.

[9] Keller, M. Innovations and perspectives of metered dose inhalers in pulmonary drug delivery, *Intl. J. Pharm.*, *186*(199), 81−90.

[10] Kaparissides, C., Alexandridou, S., Kotti, K. & Chaitidou, S. (2006). *Recent Advances in Novel Drug Delivery Systems*, azojono.

[11] Luessen, H. & Tytonis, B. V. (2007). Pulmonary drug delivery new perspectives on inhalers and inhalables, *On Drug Delivery.*

[12] Callion, O. N. M. (1996). Jet nebulizer for pulmonary drug delivery, *Intl. J. Pharm.*, 1−11.

[13] Le Brun, P. P. H., de Boer, A. H., Heinemann, H. G. M. & Frijlink, H. W. (2000). A review of the technical aspects of drug nebulization, Kluwer Academic Publishers, *Pharm. World Sci.*, *22*(3), 75−81.

[14] Labiris, N. R. & Dolovich, M. B. (2003). Pulmonary drug delivery. Part II: The role of inhalant delivery devices and drug formulations in therapeutic effectiveness of aerosolized medications, *Brit. J. Clin. Pharmacol.*, *56*(6), 600–612.
[15] Moran, A. (2007). Astec Project Ltd, Next generation of automated pulmonary drug delivery system, *On Drug Delivery*.
[16] Li, H-Y. & Seville, P. C. (2009). Novel PMDI formulations for pulmonary delivery of proteins, *Intl. J. Pharm.*, 1–6.
[17] Senior, K. (2000). New technology for pulmonary drug delivery, *PSTT*, Vol. *3*, No. 8.
[18] Hiller, F. C. (2004). Therapeutic aerosols: An overview from a clinical perspective, Modern Pharmaceutics Marcel Dekker, 1–31.

*Chapter 10*

# THE NOVEL DRUG DELIVERY SYSTEMS FOR CANCER THERAPY

Most anti-cancer drugs are based on a differential for killing cancer cells. Development of cancer therapy has largely been driven by the assumption that if we screen enough drugs we can find a drug with the precise mechanism of action that will allow for a greater discrimination between normal cells and cancer cells. Although, from a biochemical (or pharmacological) viewpoint – there is very little difference between normal cells and cancer cells, *i.e.*, pharmacological specificity of anti-cancer agents are very low for most cancers. Besides their side effects are usually dosage limiting. Conventional formulations have been optimized by: careful timing, slow infusions, monitoring of serum drug levels, and multi-drug regimes, etc.

## pH-DEPENDENT GLYCOL CHITOSAN AND DEXTRAN SULFATE NANOPARTICLES FOR EFFECTIVE BRAIN CANCER TREATMENT

Chitosan is a product obtained from the de-N-acetylation of chitin, a co-polymer of N-acetyl-glucosamine and N-glucosamine units linked by ß-(1, 4) glycosidic bonds, in the presence of hot alkali (Figure 1).

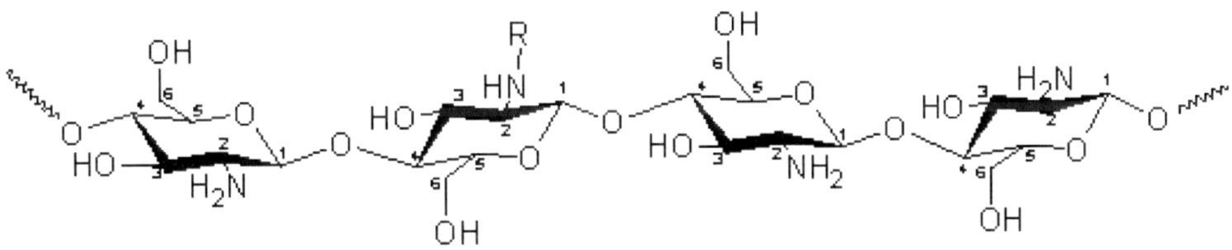


R=H or COCH3

This process causes changes in molecular weight and the degree of deacetylation of the product. The weight average molecular weight of chitin is $1.03 \times 10^6$ to $2.5 \times 10^6$, but the N-deacetylation reaction reduces this to $1 \times 10^5$ to $5 \times 10^5$. Depending on the source and preparation procedure, its molecular weight may range from 300 to over 1000 kDa with a degree of deacetylation from 30 to 95%. Thus, chitosan is a collective name representing a family of de-N-acetylated chitins with different degrees of deacetylation. In general, when the number of N-acetyl-glucosamine units is higher than 50%, the biopolymer is termed chitin. Conversely, when the number of N-glucosamine units is higher, the term chitosan is used. Chitosan has three types of reactive functional groups, an amino group as well as both primary and secondary hydroxyl groups at the C-2, C-3, and C-6 positions, respectively. These groups allow chemical modifications of chitosan that include acylation, N-phthaloylation, tosylation, alkylation, Schiff base formation, reductive alkylation, O-carboxymethylation, N-carboxyalkylation, silylation, and graft copolymerization. Chemical modifications of these groups resulted in the production of numerous useful derivatives for specific application. Applications of chitin are limited compared to chitosan because chitin is chemically inert and is insoluble in both water and acid, while chitosan is relatively reactive and can be produced in various forms. Chitosan is normally insoluble in neutral or basic pH conditions, while soluble in acidic pH. The solubility of chitosan depends upon the distribution of free amino and N-acetyl groups. In dilute acids (pH< 6), the free amino groups are protonated and the molecule becomes soluble. Chitosan and its derivatives possess some special properties for use in regenerative medicine. Several studies have examined the host tissue response to chitosan based implants. In general, these materials are non-toxic and biodegradable with living tissues, evoke a minimal foreign body reaction with little or no fibrous encapsulation. Lysozyme is the primary enzyme responsible for *in vivo* degradation of chitosan through hydrolysis of acetylated residues. The degradation rate of chitosan is inversely related to the degree of crystallinity, and thus on deacetylation. Chitosan have cationic nature that is primarily responsible for electrostatic interactions with anionic glycosaminoglycans (GAG), proteoglycans and other negatively charged molecules. This property allows chitosan to retain and concentrate biomolecules since a large number of cytokines/growth factors are linked to GAG. For drug delivery applications, CS and CS derivatives have been extensively investigated because of their biocompatibility, non-toxicity, and bioadhesive properties. GCS is a chitosan derivative conjugated with ethylene glycol branches, water soluble at neutral/acidic pH values and viscose. Modified glycol chitosan nanoparticles for cancer therapy have been recently formulated for the administration of docetaxel camptothecin. An important choice of dextran as the molecular backbone was based on practicality and availability. Dextran sulfate sodium is a long-chain polymer of glucose containing 17-20% sulfur. It has been used as an anticoagulant and also has been shown to inhibit the binding of HIV-1 to CD4-POSITIVE T-LYMPHOCYTES. It is commonly used as both an experimental and clinical laboratory reagent and has been investigated for use as an antiviral agent, in the treatment of hypolipidemia, and for the prevention of free radical damage, among other applications. In this context, the aim of the present work is to develop and characterize new modified chitosan nanoparticles intended as vectors of the antifolic agent methotrexate (MTX). From a theraoeutic viewpoint, MTX is substrate of drug efflux transporters located at the apical side of Blood – Brain – Barrier (BBB) such as p-glycoprotein, multidrug resistance protein and breast cancer protein. Thus, the MTX clinical efficacy for the treatment of brain tumors is limited. Therefore, the nanocapsulation of the drug in the polymeric networks of Glycol chitosan network was aimed to mask the drug from the attacks of the aforementioned efflux systems and moreover, to overcome the BBB, the role of polymeric network was investigated.

# BIODEGRADABLE MODIFIED CHITOSAN NANOPARTICLES FOR PHOTODYNAMIC TREATMENT OF CANCER

Photodynamic therapy (PDT) is a two-step treatment used to help treat symptoms caused by cancer or its treatment. It is called two-step since both a drug and laser treatment are used. PDT is most often used to help people who have cancer of the esophagus or lung. This treatment is given while you are in the hospital. Biodegradable nanoparticles have received much attention as possible drug-delivering agents. Their main advantages lie in high drug loading, the possibility of controlling the drug release, and a large variety of materials and manufacturing processes.

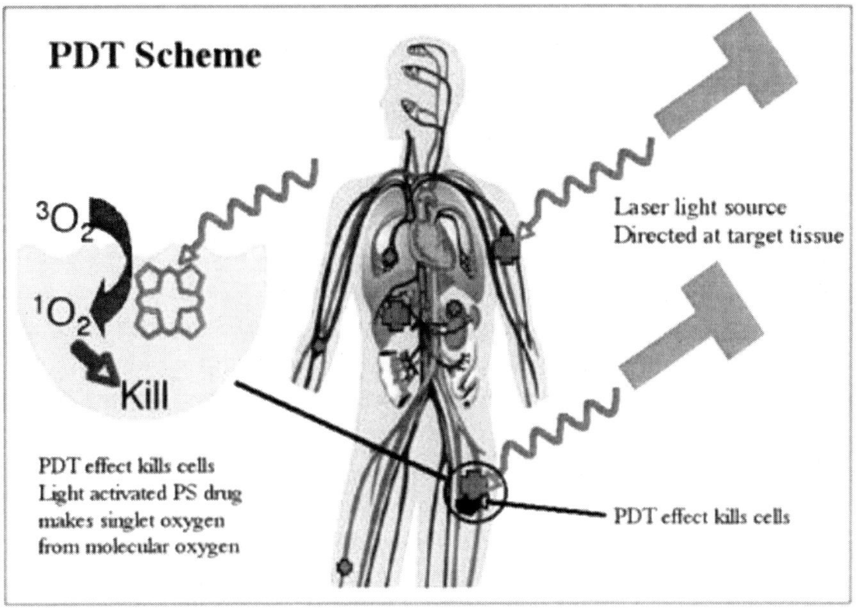


Biocompatible polymers are predominantly favored over small molecules for drug delivery systems because of the opportunities that polymers offer in achieving optimal medication effects. The chemical composition and architecture of polymers can be readily tailored to accommodate drugs with varying degrees of hydrophobicity, molecular weight, pH, etc. Furthermore, the surface properties, morphologies, and composition of polymer matrices can be easily optimized to achieve the desired biocompatibility, controlled polymer degradation and drug release kinetics. Recently, studies in PDT have thus focused on the use of polymer-based nanoparticles for PS delivery and release systems, in particular those with biocompatible and biodegradable polymers. For example, the surface of particles can be modified with poly(ethylene oxide) to improve the carrier's biocompatibility and biodistribution. Different strategies have been developed for the preparation of polymer-based nanomaterials for the delivery of PS molecules in PDT. Among these methodologies, the common method is the emulsion polymerization or the interfacial polymerization in water-in-oil or oil-in-water microemulsions. The resulting nanomaterials can be either nanospheres or nanocapsules (core-shell nanoparticles) depending upon the synthetic approach among the various PSs investigated for PDT, phthalocyanines have been found to be highly promising.

However, one of the problems in using phthalocyanines for PDT arises from their intrinsic tendency toward aggregation. This intermolecular association process promotes the non-radiative internal conversion, almost invariably causing a shortening of the triplet lifetime and a drastic reduction in the overall efficiency of singlet oxygen production. As early as 1991, the preparation of tetrasulfonated zinc phthalocyanine (ZnPcS4) or aluminium naphthalocyanine entrapped into poly (isobutylcyanoacrylate) or poly (ethylbutylcyano-acrylate) nanocapsules or nanospheres by interfacial polymerization in water-in-oil emulsion was reported. The prepared nanocapsule sizes ranged from 150 to 250 nm, while the nanospheres had a broader size distribution from 10 to 380 nm. The PS encapsulation efficiency was very high and could reach 100% when prepared under optimal experimental conditions. PS binding in the nanoparticles was between 75 and 80%. In a subsequent study with cultured V-79 hamster cells as a model, it was found that the nanoencapsulation enhanced the cellular localization of the dyes providing a high level of *in vitro* phototoxic activity. It was found that the formulation of the biodegradable nanoparticle improved the PDT response of the tumor while providing prolonged tumor sensitivity towards PDT. Recently, a similar method has been applied to the preparation of poly (D,L-lactide-co-glycolide) (50:50 and 75:25 PLGA) and poly(D,L-lactide) nanoparticles entrapped with the PS meso-tetra(hydroxyphenyl)porphyrin (p-THPP). The resulting nanoparticles have an average particle size of 130 nm and the PS loading was improved to near 8%. The cellular uptake, localization and phototoxicity of PS-loaded carriers have been studied in macrophage-like J774 cells and HT 29 human adenocarcinoma cells. It was found that the cellular uptake was reduced with mTHPC encapsulated into surface-modified nanocapsules – PLA-PEG and polox PLA – compared with naked PLA. However, the phototoxicity was much less affected and a protecting effect against plasma proteins was observed. As explained earlier, singlet oxygen is the main cytotoxic species in PDT, and hence a fundamental requirement for a PS carrier is that it should not significantly quench singlet oxygen. In this research paper, we have been synthesized a biodegradable and biocompatible polymeric nanoparticles for PS delivery and release system. This system is able to target different organs and control the release of the PS molecules by the incorporation of site-specific moieties. The system is based on a modified chitosan – PS conjugate that uses for PDT.

## A Novel Superparamagnetic Chitosan Hydrogel/Chlorin E6 Nanocarriers

Photodynamic therapy (PDT) of cancer is a relatively new method of treatment whereby non-toxic drugs (sensitizers) and non-hazardous visible and near infrared light (VIS/NIR) combine to generate cytotoxic reactive oxygen species (ROS) at a selected treatment site. The application of PDT in tumor care has been motivated by the quest for a treatment that is at least as effective, but more selective than radio and chemo-therapies; thus minimizing side effects. Current PDT aims at directly killing tumor cells and typically consists of five steps. 1. Administration of a photosensitizer, usually intravenously (i.v.). 2. A delay period that allows for retention or accumulation of photosensitizers in the target tissue. 3. Illumination of the target tissue (transcutanously or interstitially via optical fibers for inner organs) with consequent local generation of cytotoxic ROS. 4. Development of tumor necrosis and tumor eradication. 5. Tissue remodeling and healing.

New sensitizers have usually been designed to achieve preferential accumulation in the tumor cells while photoexcitation techniques have been developed to irradiate relatively deep tumor tissue thus expanding the usefulness and efficacy of PDT. Different porphyrinoids have been suggested as the preferred PDT sensitizers, although cyclic and long-chain polyenes with a significant light absorption in the UV-VIS such as indocyanines and hypericin were also considered. For several decades, research and pre-clinical efforts have focused on hematoporphyrin derivatives (HpD) and other peripherally substituted porphyrins of (near) D4h symmetry. The preferential accumulation of hematoporphyrin derivatives in tumors relative to normal tissues, the formation of cytotoxic ROS, and necrotic or apoptotic processes that culminate in tumor eradication, were promising in early PDT development; however, none of the clinically or pre-clinically used sensitizers have yet shown sufficiently high accumulation in the malignant tumor cells to enable optimally-selective and efficient treatment. In addition, the high attenuation of light in the UV-VIS domain, required for activation of tetrapyrrole sensitizers with D4h-symmetry within animal tissues, has prevented treatment of massive tumors. Thus, the current clinical targets of PDT include relatively shallow malignant and benign tumors, choroidal neovascularization (CNV) in age-related macular degeneration (AMD), atherosclerotic lesions, as well as bacterial and viral infections. In addition, the photodynamic effect can be used as a subsidiary but method selectively for light-enhanced delivery of a drug to the body area/organ to be treated. Over the past decade, the use of PDT has increased significantly, mainly in cancer and AMD treatment. Protoporphyrin-based photosensitizers (trade names: Photofrin, Photosan, Photoheme, HpD, Levulan, Visudyne) have been approved for clinical use, and successfully employed in PDT in many countries. Chn E6, formed by anaerobic alkaline hydrolysis of Pheide *a*, is a tricarboxylic water-soluble derivative absorbing light at 654 nm and has an $\varphi\Delta$ of about 0.7. [Sn]IV-Chn e6 absorbs at 632 nm and generates singlet oxygen with $\varphi\Delta$ equal to 0.83 (Amides of Chn E6 at C-131 were regioselectively obtained by aminolysis of the isocyclic ring. Mono-Laspartyl Chn E*6* (code name NP*e*6), obtained by amidation of the acetic acid residue of Chn E*6* with L-aspartic acid, possessed an increased hydrophilicity and a $\varphi\Delta$ of 0.77. Targeted delivery of [Sn]IV-Chn E6, conjugated to epidermal growth factor (EGF) through a carrier, showed increased phototoxicity to squamous carcinoma cells in relation to phototoxicity of the non-conjugated sensitizer, which express an increased number of EGF-receptors; however, the affinity of the conjugate was strongly dependent on the carrier used (dextran, polyvinyl alcohol, or human serum albumin). [Sn]IV-Chn E6 linked to IgG selectively killed otherwise resistant strains of *Staphylococcus aureus*. The importance of positive charges on the sensitizer's periphery to the overall photodynamic activity was demonstrated by the killing of both gram-positive and gram-negative bacteria by cationic conjugates of Chn E6. Conjugates for doubly-targeted delivery to both cell and nucleus were constructed using Chn E6 together with BSA as a carrier, insulin as internalizable ligand and viral T-antigen within a β-galactosidase fusion protein as a nuclear localization factor. This molecular assembly demonstrated a 2400-fold higher photodynamic activity in human hepatoma cells (EC50 of 0.13 nM) compared to free Chn E6. Co-incubation with adenovirus additionally increased the nuclear photosensitizing activity.

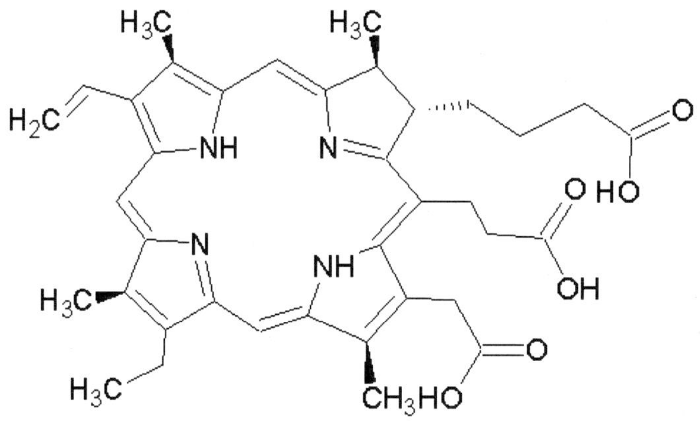

## CHLORIN E6

Chn E6 was also attached to linear or branched peptides for guidance to both cytoplasmic and nuclear targets: such conjugates displayed 400- and 40-fold more phototoxicity in CHO and RIF-1 cells than Chn E6 alone. A new type of chitosan (CS) hydrogel using dextran sulfate (DS) as a polyanionic polymer was developed to achieve complex coacervation for the incorporation and controlled release of an anti-angiogenesis hexapeptide, this was the first report describing the use of DS to formulate CS based hydrogels.

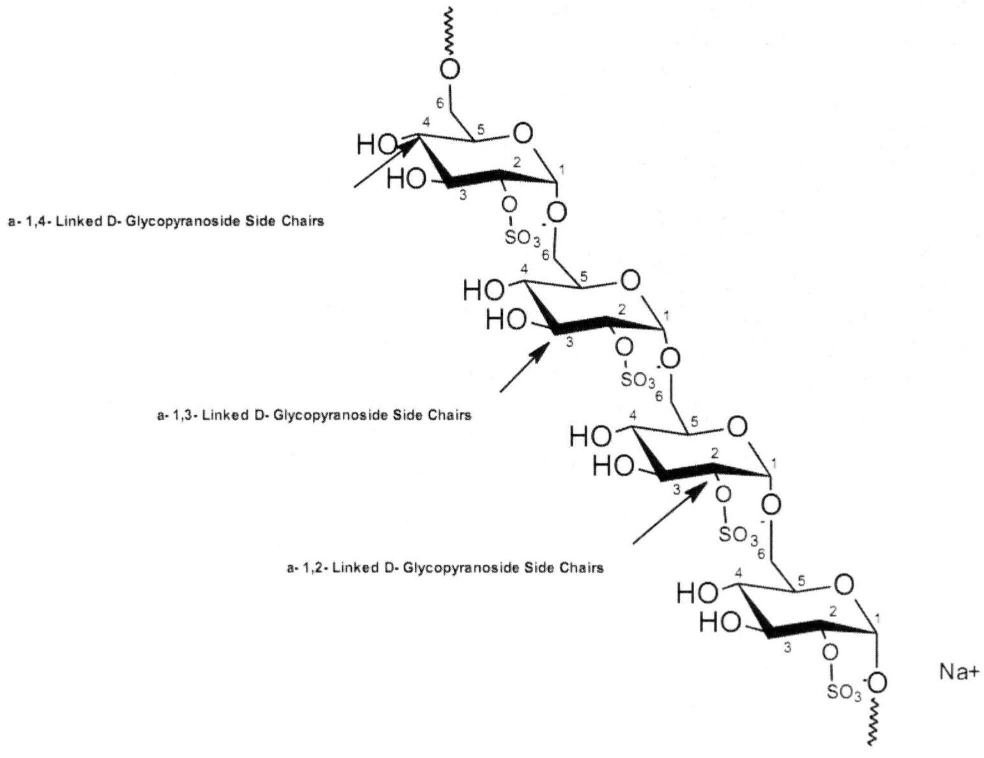

## DEXTRAN SULFATE SODIUM

Although there have been investigations of how the properties of CS and formulation variables such as CS molecular weight (MW), concentrations of CS and Chn E6, and formulation pH affect the formation and encapsulation capability of hydrogels; to our knowledge, no attempts have been made to study how the ratio of CS to the oppositely charged polymer influences the formulation and properties of hydrogel. The system is based on a CS-DS/Chn E6 conjugate that uses for PDT. These hydrogels had good spherical geometry. It is obvious that the surface of the hydrogel shrank and a densely cross-linked gel structure was formed. This may explain the greater retardation of drug release from matrices of a higher cross-linked content. The average drug entrapment was found to be 90.34±0.32% in the hydrogel. FESEM images of super paramagnetic CS–DS hydrogel (Figure 1) show that hydrogel have a solid and near consistent structure.

Furthermore, the incorporation of Chn E6 into the hydrogel produced a smooth surface and compact structure. The particle size observed in FESEM is smaller than that measured by the Zetasizer. This is because dried hydrogel was used in FESEM, whereas for particles in the liquid dispersion was analyzed by the Zetasizer. CS–DS hydrogel are hydrophilic and would be expected to swell in water, thus producing a large hydrodynamic size when measured by the Zetasizer. Results of drug content and EE demonstrated that drug content increased from 9.65±0.50 mg/100mg to 29.38±0.43 mg/100mg of hydrogel with increasing the amount of drug from 5 to 20% (w/w). No signification increase in drug content was observed when the amount of drug was increased further, *i.e.*, above 15% (w/w), which could be due to the limited solubility of the drug in DMF and that is endorsed from the presence of drug particles on the surface of the hydrogel prepared with 20% of drug concentration. The percent EE was increased up to 90.34±0.20% when increasing polymer concentrations to 4%. Concentration of the cross-linking agent exhibited no significant effect on percent EE. Chn E6-loaded hydrogel was obtained spontaneously upon the mixing of the DS aqueous solution (0.1%, w/v) with the CS solution (0.1%, w/v) under magnetic stirring, with Chlorin E6 dissolved in CS–DS solution.

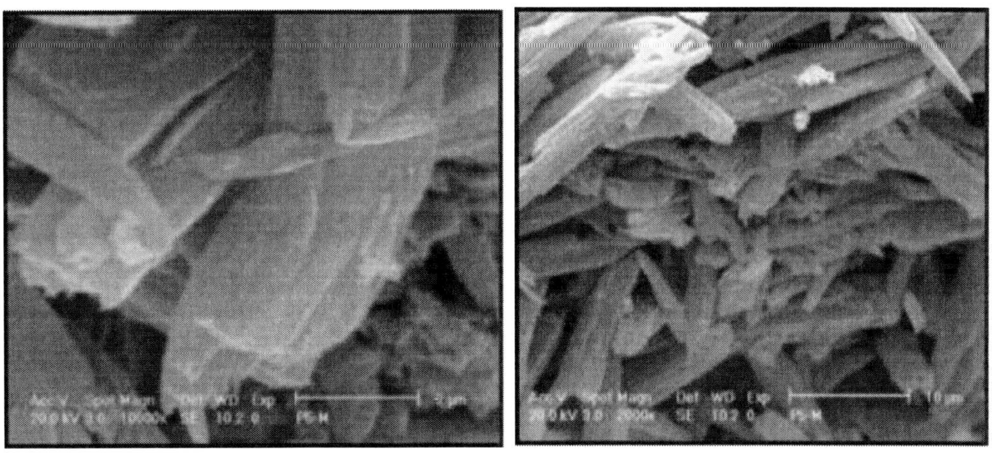

Figure 1. Morphology of chn E6 loaded CS-DS hydrogel by emission scanning electron microscopy field.

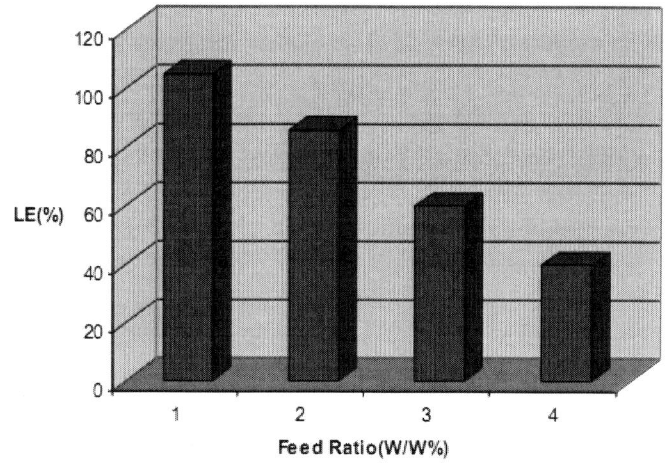


(a)

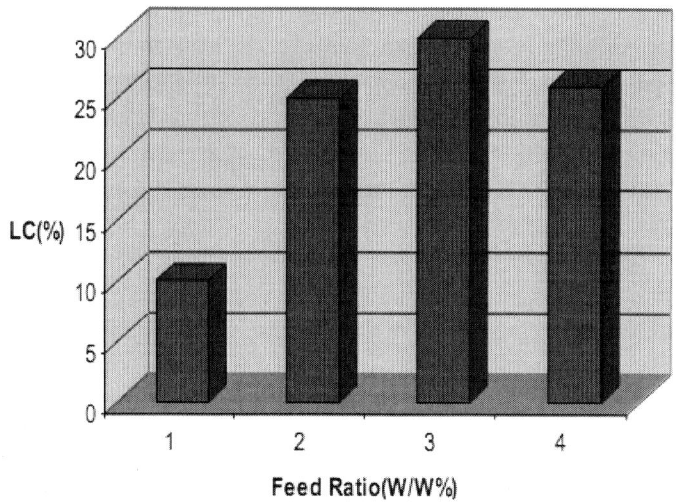


(b)

Figure 2. (a) Loading efficiency and (b) loading capacity of CS-DS hydrogel with Chn E6.

The incorporation of Chn E6 into the CS–DS hydrogel resulted in a sharp increase in the particle size of the nanoparticle dispersion. The significant increases in particle size provided a good induction for the incorporation of Chn E6 into a CS-DS hydrogel. A study was undertaken to investigate the order of effect that Chn E6 has mixing with CS and DS. The data obtained showed that the order of Chn E6 mixing had no effect on the size, entrapment efficiency, and yield of the Chn E6-loaded hydrogel. The loading efficiency and loading capacity of the nanoparticle dispersions are presented in Figures 2a and 2b, respectively. It was shown that the Chn E6 /polymer feed ratios were up to 30%(w/w), the loading efficiency was almost quantitative, resulting in loading capacities were up to 30% and the Chn E6 concentration in the dispersion of 1.8 mg/mL; which is markedly higher than the aqueous solubility of Chn E6. At 10% (w/w) loading, the dispersions stayed clear for more than a week. At higher loading of 20% and 30% (w/w), a precipitate was observed in the dispersions for the loaded hydrogel within 1 day after preparation, indicating that the loading capacity of these hydrogels was exceeded. In contrast, thc hydrogel composed of polymer remained clear for more than a week at 20% and 2 days for 30% loading, which demonstrates that these end groups have a favorable effect on the loading and solubilization capacity of the hydrogel. The *in vitro* photocytotoxicity effect of Chn E6 loaded hydrogel on 14C cells was determined after 6 h incubation and compared with free Chn E6. Either various dilutions of loaded nanoparticles at a fixed Chn E6 /polymer ratio of 15%(w/w) were used, or the polymer concentration was fixed at 0.5 mg/mL and the Chn E6 loading was varied between 0.004% and 0.32%(w/w) in order to secure a polymer concentration of 100 times above its critical aggregation concentration. The results demonstrate that the effect on the viability of 14C cells of Chn E6 – polymer hydrogel at a fixed Chn E6/polymer ratio is comparable to that of free Chn E6. At a fixed high polymer concentration, the nanoparticles with various loadings were hardly toxic to cells even at the highest Chn E6 concentration used, as indicated by a cell viability of 80%.

Loaded hydrogel and free Chn E6 were diluted in medium to a concentration of 12 μM Chn and incubated with 14C cells for 6 h to determine the cellular uptake of Chn E6. The 15% (w/w) Chn loaded and polymer nanoparticles showed a lower cellular uptake of Chn E6 compared to free Chn E6, by a factor of 2.8 and 3.7, respectively. More importantly, hardly any Chn E6 uptake was measured after incubation with the hydrogel yielding a cellular uptake comparable of free Chn E6. The results of the photocytotoxicity and cellular uptake of the different formulations suggest that when the concentration of polymer is kept well above, Chn E6 remains stably encapsulated in the hydrogel even in the presence of serum, and that the intact nanoparticles are not taken up by the cells, which may be ascribed to the presence of a dense polymer-layer. Only after dilution of the Chn E6 -loaded hydrogel to polymer concentrations below and after hydrogel degradation, the released chn E6 can be taken up by 14C cells and exert its cytotoxic effect upon illumination.

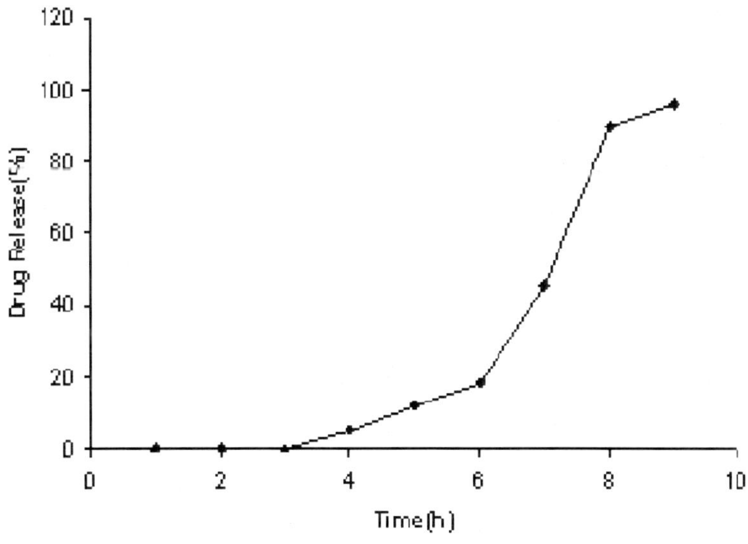


Figure 3. Release data of the Chn E6 from CS-DS hydrogel at 37°C.

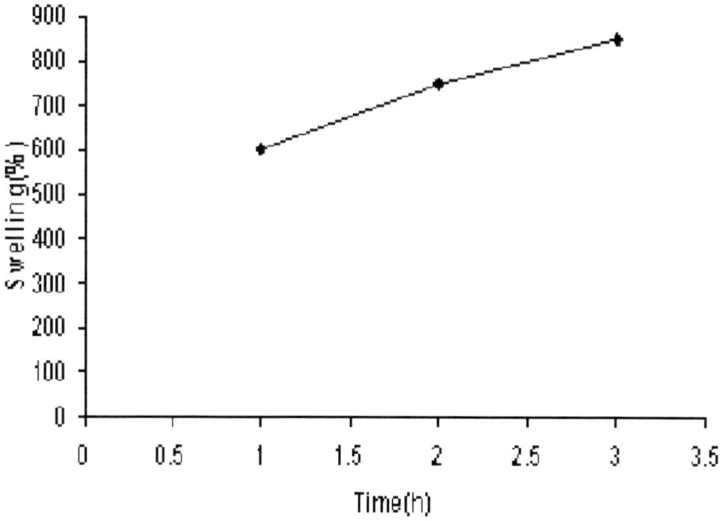


Figure 4. Time-dependent swelling behavior of Chn E6 loaded CS-DS hydrogel at 37°C.

**Table 2. Conductivity of dispersion of hydrogel in the different pH**

| Weight ratio of CS:DS | pH of hydrogel | Conductivity of dispersion (in mυ) |
|---|---|---|
| 5:4.5 | 4.0 | 3.89 |
| 5:5 | 3.9 | 3.94 |
| 5:8.5 | 4.1 | 3.99 |

A nanopolymer bonded Chn E6 (50 mg) was poured into a 3 mL of aqueous buffer solution (pH 7.4) (Figure 3). The mixture was introduced into a cellophane membrane dialysis bag. The bag was closed and transferred to a flask containing 20 mL of the same solution maintained at 37° C.

The external solution was continuously stirred, and 3 mL samples were removed at selected intervals. The triplicate samples were analyzed by UV spectrophotometer, and the quantity of Chn E6 was determined by using a standard calibration curve obtained under the same conditions. The primary mechanism for the release of Chn E6 from these matrix systems *in vitro* are swelling, diffusion, and disintegration (Figure 4).

*In vitro* degradation of CS-DS hydrogels were prepared by solution casing method occurred less rapidly as the degree 73% deacetylated showed slower biodegradation. Since the grade of CS-DS used in the present study was of high molecular weight with a degree of deacetylation ≥85%, significant retardation of release of Chn E6 from bead is attributed to the polymer characteristics. In addition, diffusion of Chn E6 may have been hindered by increased tortuosity of polymer accompanied by a swelling mechanism. The opposite charges of CS-DS hydrogel were responsible for the formation of nanoparticles. The charge ration between the negatively charged sulfate groups (N) in DS and the positively charged amine groups (P) in CS was determined. Under the experimental conditions (pH 3–4), DS carries ~74 sulfate groups per mole, equivalent to 5.78×10–3 negatively charged groups per gram of

DS; CS has ~2073 amino groups per gram of CS. Table 2 shows conductivity of dispersion of hydrogel.

## A Novel Modified Starch Hydrogels for Photodynamic Treatment of Cancer

PDT is based on the fact that some photosensitizers (PSs) can accumulate a higher concentration in tumor cells than in healthy cells upon systemic administration. By matching the wavelength of the therapeutic light to the absorption peak of the sensitizers, most of the light is absorbed by the PSs, and the excited PS molecules can then transfer their energy to surrounding oxygen molecules, which are normally in their triplet ground state. These results are in the formation of reactive oxygen species (ROS) such as singlet oxygen ($^1O_2$) or free radicals. These ROSs are responsible for oxidizing various cellular compartments, resulting in interversible damage to tumor cells. PSs, oxygen, and light are three crucial components for this photon-induced toxicity effect. Most exiting PSs are hydrophobic and may aggregate easily in a biological environment. Even for hydrophilic PSs, the selective accumulation in tumor tissues is not high enough for clinical use. To overcome these limitations, colloidal carriers for PSs, such as liposomes and polymeric micelles, have been investigated. This study as described in this paper, carboxymethylstarch(CMS)-Dextran sulfate(DS) hydrogels with encapsulated mTHPP, which has been widely investigated and officially approved for use in clinical treatments were synthesized and utilized as drug carriers.

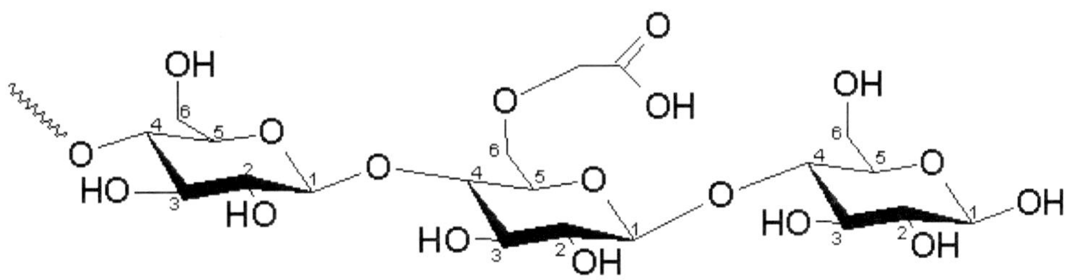

CARBOXYMETHYL STARCH

The PS-encapsulated CMS-DS hydrogels can be easily prepared with desired size, shape, and porosity, and are very stable. They are more biocompatible than other nanomaterials, and they can effectively protect the PSs against denaturation induced by the extreme bioenvironment. The CMS-DS hydrogel encapsulation is relatively transparent for both the activated light and signal light. To accumulate PSs selectively in tumor tissues and reduce the dosage of PSs, their surfaces can be functionalized with various groups and further conjugated with antibodies for specific uptake in an *in vivo* experiment. An investigation showed how CMS-DS hydrogels were used with encapsulated mTHPP for PDT treatment. The CMS-DS hydrogels were synthesized, characterized, and utilized for *in vitro* imaging of tumor cells. The effect of photon-induced toxicity on hydrogels was also demonstrated after performing a comparison with some control experiments. The mTHPP-encapsulated CMS-DS hydrogels showed good potential applications to, e.g., photodynamic therapy for cancers.

One potential challenge of PDT therapy is that many PS agents are lipophilic, making parenteral administration problematic. In addition, systemic administration of a PS leads to generalized photosensitivity and the temporary need to avoid light exposure. Various strategies to overcome these limitations have been investigated, including conjugation of PS agent to water soluble polymers and colloidal administration, as well as encapsulation in hydrogel carriers such as micelles. Recent progress has been made in the design of polymeric hydrogels for nanoscale therapeutic and diagnostic applications. Polymeric hydrogels are composed of amphiphilic block copolymers that contain distinguished hydrophobic and hydrophilic segments. The distinct chemical nature of the two blocks resulted in thermodynamic phase separation in aqueous solution and formation of nanoscale supramolecular structures. A new type of carboxymethyl Starch (CMS) hydrogel using dextran sulfate (DS) as a polyanionic polymer was developed to achieve complex coacervation for the incorporation and controlled release of an anti-angiogenesis hexapeptide, this was the first report describing the use of DS to formulate CMS based hydrogels.

Although there have been investigations of how the properties of CMS and formulation variables such as CMS molecular weight ($M_W$), concentrations of CMS and mTHPP, and formulation pH affect the formation and encapsulation capability of hydrogels; to our knowledge, no attempts have been made to study how the ratio of CMS to the oppositely charged polymer influences the formulation and properties of hydrogel. This system is able to target different organs and control the release of the PS molecules by the incorporation of site-specific moieties.

The system is based on a CMS-DS/mTHPP conjugate that uses for PDT. Many functional properties of CMS-DS such as paste and gel storage stability have significantly improved. The studies on the effects in the degree of starch substitution for the physicochemical and properties of CMS-DS hydrogel have revealed that the degree of substitution in hydrogel increased from 0.10 to 0.20 both viscosity and freeze-thaw stability of hydrogel decreased. The amylase content of starch influences many physicochemical and functional properties of starch products. It is also worthwhile to investigate whether amylase content plays a role on the properties of hydrogel with similar degree of substitution level.

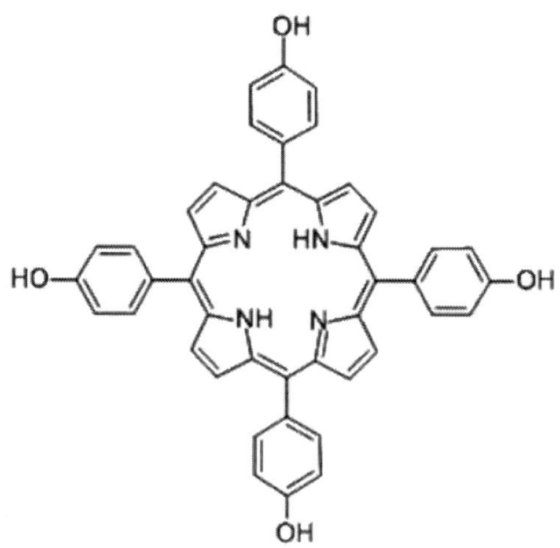


meso-tetra(hydroxyphenyl)porphyrin (p-THPP)

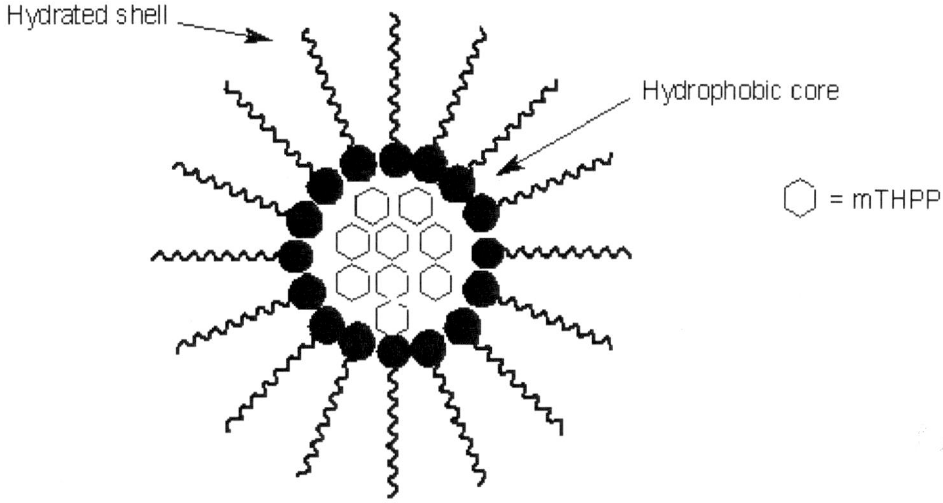


Scheme 1. The chemical structures are shown for CMS-DS hydrogels and the mTHPP Morphology of hydrogels.

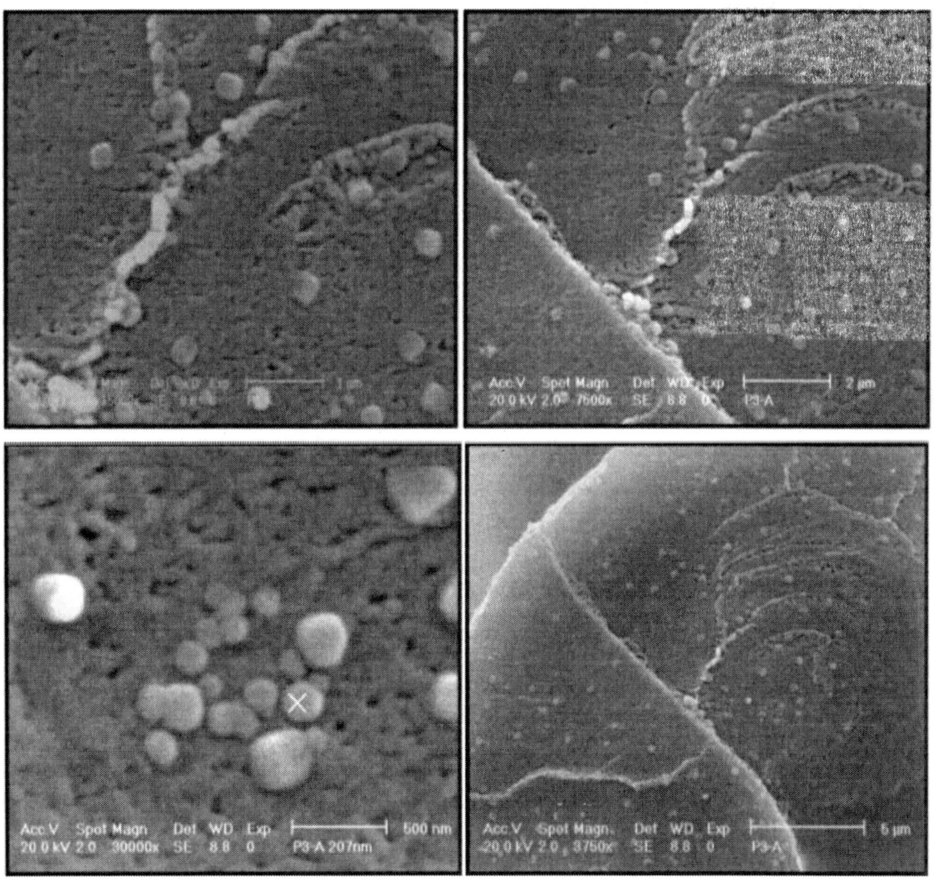


Figure 5. (Continued).

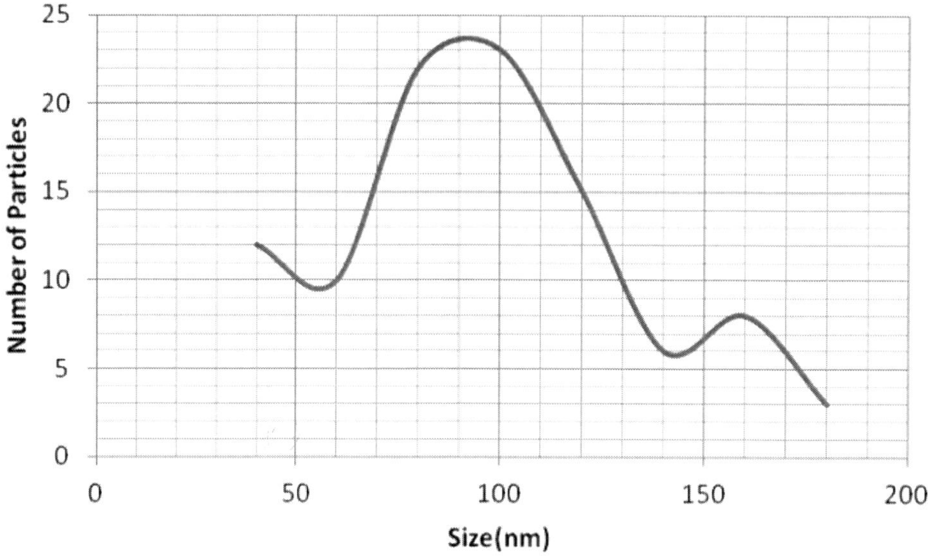


Figure 5. a). Morphology of mTHPP loaded CMS-DS hydrogel by emission scanning electron microscopy field. b). The size distribution of m-THPP /CMS-DS hydrogels was determined from the SEM picture.

The mTHPP loaded CMS-DS hydrogels were characterized by the FESEM for their size and distribution. Data showed that hydrogel have a solid and near consistent structure. These hydrogels have good spherical geometry. Figure 5a shows a representative SEM image of mTHPP hydrogels. Also, the results show that the surface of CMS-DS hydrogels shrank and a densely cross-linked gel structure was formed. There was a retardation of the drug release from matrices with a higher cross-linked content. The average drug entrapment was found to be 92.18±0.10 % in the hydrogel.

**Table 3. Compositions and characteristics of different mTHPP loaded CMS-DS hydrogels**

| Variables | Values | Degree of swelling (%) | Average particle size (nm) | Encapsulation efficiency(%) |
|---|---|---|---|---|
| mTHPP | 0 | 1.64±0.10 | 1.76±0.12 | - |
| | 5 | 1.34±0.03 | 2.00±0.10 | 73.41±0.12 |
| | 10 | 0.86±0.12 | 2.45±0.04 | 80.10±0.10 |
| | 15 | 0.68±0.05 | 2.43±0.80 | 83.64±0.05 |
| | 20 | 0.49±0.08 | 3.00±0.06 | 92.18±0.08 |
| Time(h) | 10 min | 0.45±0.34 | 1.74±0.13 | 65.56±0.24 |
| | 20 min | 0.67±0.12 | 1.62±0.10 | 87.12±0.10 |
| | 40 min | 0.87±0.03 | 1.43±0.04 | 75.49±0.02 |
| Drying(°C) | Lyophilization | 0.74±0.06 | 1.15±0.16 | 68.10±0.03 |
| | 45 | 0.61±0.10 | 1.24±0.04 | 57.12±0.12 |

**Table 4. mTHPP hydrogel fabrication procedure with resulting yield and drug loading parameter**

| Theoretical loading (%) | Hydrogel Yield(%) | mTHPP Yield (%) | Net loading Efficiency (%) |
|---|---|---|---|
| 5 | 91.0±1.0 | 89.3±0.03 | 99.2± 1.2 |
| 10 | 93.4±1.2 | 90.5±0.01 | 100.4±1.4 |
| 20 | 94.1±1.5 | 85.6±0.08 | 84.5±1.5 |

The size of CMS-DS hydrogels was estimated by scanning electron microscopy (SEM) of the dried hydrogel dispersions. The determination of hydrogel size by SEM under a dry state does not result in an accurate absolute value of the hydrated hydrogel size in dispersion, but only visualizes size range and particle shape.

The SEM picture of m-THPP-hydrogels is shown in Figure 5b. SEM analyses confirmed the relatively broad size distribution of the CMS-DS hydrogels, ranging from about 40-100 nm. The incorporation of mTHPP into the CMS-DS hydrogel produced a smooth surface and compact structure. The CMS-DS hydrogels are hydrophilic and would be expected to swell in water, thus producing a large hydrodynamic size when measured by the Zetasizer. The particle size of hydrogels is measured (Table 3).

## Drug Content and Encapsulation Efficiency

Loading content of mTHPP hydrogels with different theoretical loading densities were analyzed, and the results are shown in Table 4. Drug loading density, loading efficiency, and mTHPP and hydrogel yields of the hydrogel with 5, 10, 20% theoretical loading densities were calculated. Results show that mTHPP was effectively loaded into CMS-DS hydrogels, with high loading efficiency in all formulations. For mTHPP, loading densities for 5, 10 and 20% theoretical loading were 4.6±0.1, 9.8±0.2, 16.8±0.3%, respectively. Net loading efficiencies of 5, 10% theoretical loading were quantitative, whereas that of 20% theoretical loading was 89.26±1.5%. The mTHPP and hydrogel yields were also sufficiently high to confirm hydrogel encapsulation of mTHPP as an effective strategy for solubilization of PS.

## Incorporation of mTHPP in Hydrogel

The incorporation of mTHPP in the CMS–DS hydrogel resulted in a sharp increase in the particle size of the nanoparticle dispersion. The mTHPP-loaded hydrogel was obtained spontaneously upon the mixing of the DS aqueous solution (0.1%, w/v) with the CMS solution (0.1%, w/v) under magnetic stirring, with mTHPP dissolved in CMS–DS solution. The incorporation of mTHPP into the CMS–DS hydrogel resulted in a sharp increase in the particle size of the nanoparticle dispersion. The significant increases in particle size give a good induction for the incorporation of mTHPP into CMS-DS hydrogel. A study was undertaken to investigate the effect in the order of mTHPP mixing with CMS and DS. The data obtained show that the order of mTHPP mixing had no effect on the size, entrapment efficiency, and yield of mTHPP-loaded hydrogel.

## Loading of Hydrogel with mTHPP

The loading efficiency and loading capacity of the nanoparticle dispersions are presented in Figure 6, respectively. It was shown that mTHPP /polymer feed ratios of up to 20%(w/w), the loading efficiency was almost quantitative, resulting in loading capacities of up to 20% and mTHPP concentration in the dispersion of 1.3 mg/mL; which is markedly higher than the aqueous solubility of mTHPP. At 8% (w/w) loading, the dispersions stayed clear for more than a week. At higher loading of 15% and 20% (w/w), a precipitate was observed in the dispersions of loaded hydrogel within 1 day after preparation, indicating that the loading capacity of these hydrogels was exceeded. In contrast, the hydrogel composed of polymer remained clear for more than a week at 20% and 2 days for 30% loading, which demonstrates that these end groups have a favorable effect on the loading and solubilization capacity of the hydrogel.

### *In Vitro* Release Studies

Nanopolymer bonded mTHPP (50 mg) was poured into 3 mL of aqueous buffer solution (pH 7.4) (Figure 7). The mixture was introduced into a cellophane membrane dialysis bag. The bag was closed and transferred to a flask containing 20 mL of the same solution maintained at 37° C.

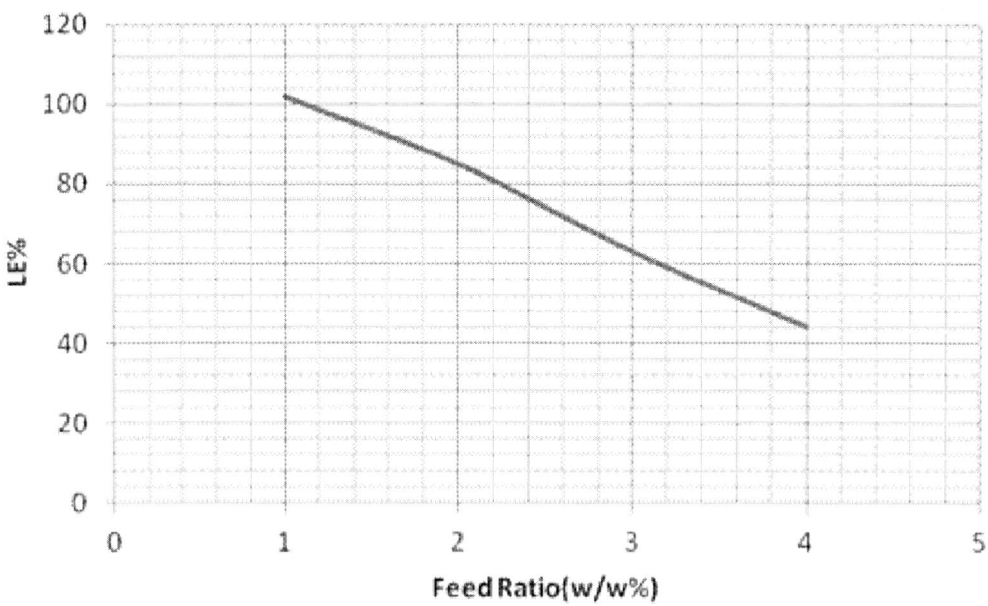


Figure 6. Loading efficiency results of CMS-DS hydrogels –mTHPP.

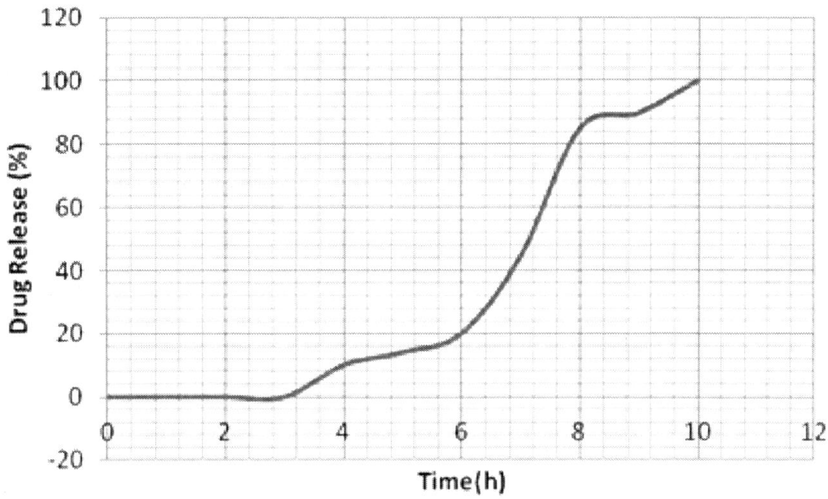


Figure 7. Release data for an mTHPP from a CMS-DS hydrogel at 37°C.

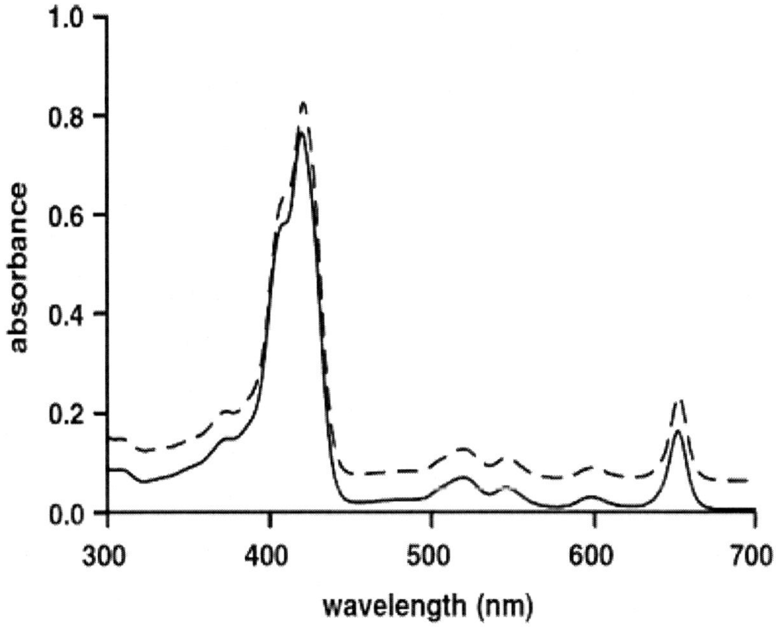


Figure 8. Absorption spectrum of mTHPP(solid line) and mTHPP loaded CMS-DS Hydrogel(dash line).

The external solution was continuously stirred, and 3 mL samples were removed at selected intervals. The triplicate samples were analyzed by UV spectrophotometer (Figure 8) and the quantity of mTHPP were determined using a standard calibration curve obtained under the same conditions. The primary mechanism release of the mTHPP from matrix systems *in vitro* were swelling, diffusion, and disintegration (Figure 9). In addition, diffusion of mTHPP may have been hindered by increased tortuosity of polymer accompanied by a swelling mechanism.

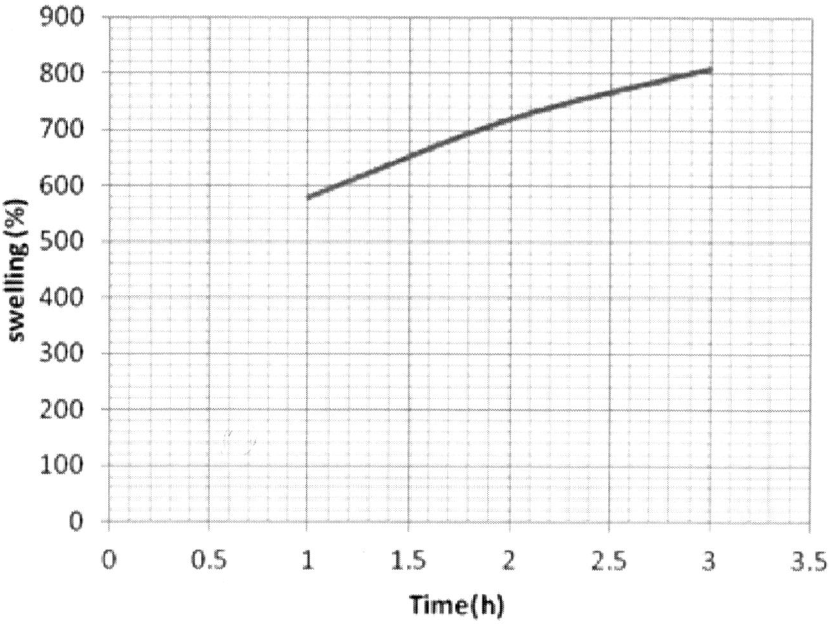


Figure 9. Time-dependent swelling behavior of mTHPP loaded CMS-DS hydrogel at 37ºC.

PDT is an effective method for cancer therapy. The efficiency with which a PS leads to the generation of $^1O_2$ and tissue damage is based on a variety of factors including triplet state quantum yield, triplet state lifetime, quantum yield of $^1O_2$, lifetime $^1O_2$, and stability of PS. The advantage of porphyrin-base PS compounds include their ability to efficiently absorb a wide range of light spectra, especially red light, which has a greater tissue penetration, as well as high quantum yield of $^1O_2$.

## References

[1] Kumar Arya, A., Kumar, L., Pokharia, D. & Tripathi, K. (2008). Applications of nanotechnology in diabetes, *Digest J. Nanomater. Biostruct.*, *3*, 221 – 225.

[2] Pison, U., Welte, T., Giersig, M. & Groneberg, D. A. (2006). Nanomedicine for respiratory diseases, *Eur. J. Pharm.*, *533*, 341-350.

[3] Pathak, P. & Katiyar, V. K. (2007). Multi-functional nanoparticles and their role in cancer drug delivery – A review. 10.2240/azojono0, *114*(3), 3−17.

[4] Weinberg, R. A. (2006). How cancer arises, *Sci. Amer.*, *275*, 62−70.

[5] Pan American Health Organization, Regional Office of the World Health Organization (WHO), *Cancer* (WHO Fact Sheet No. 297), 24/Oct/2006.

[6] Murray, C. B., Kagan, C. R. & Bawendi, M. G. (2000). *Ann. Rev. Mater. Sci. 30*, 545−610.

[7] Jain, K. (2005). Nanotechnology in clinical laboratory diagnostics: *Clin. Chim. Act. 358*, 37–54.

[8] Ferrari, M. (2005). Cancer nanotechnology: Opportunities and challenges, *Nat. Rev. Cancer*, *5*, 161– 171.

[9] Ehdaie, B. (2007). *Int. J. Biol. Sci.*, *3*, 108−110.

[10] Baker, J. R., Jr., Quintana, A., Pehlerel, L., Banazak- Hollal, M., Tomalia, D. & Raczka, E. (2001). The synthesis and testing of anti-cancer therapeutic nanodevices, *Biomedical Microdevices*, *3*(1): 61−69 .

[11] Mc Neil, S. E. (2005). Nanotechnology for biologist, *J. Leukocyte Biol.*, *78*, 585− 591.

[12] Walt,, D. R. (2002). Imaging optical sensor arrays, *Curr. Opin. Chem. Biol.*, *6*, 689–695.

[13] Nicewarner-Pena, S. R., Freeman, R. G., Reiss, B. D., He, L., Pena, D. J., et al. (2001). Submicrometer metallic barcodes, *Science*, *294*, 137–141.

[14] Cunin, F., Schmedake, T. A., Link, J. R., Li, Y. Y., Koh, J., et al. (2002). Biomolecular screening with encoded porous-silicon photonic crystals, *Nat. Mater.*, *1*, 39–41.

[15] Dejneka, M. J., Streltsov, A., Pal, S., Frutos, A. G., Powell, C. L., et al. (2003). Rare earth-doped glass microbarcodes, Proc. *Natl. Acad. Sci. USA*, *100*, 389–393.

[16] Cao, Y. W. C., Jin, R. C. & Mirkin, C. A. (2002). Nanoparticles with Raman spectroscopic fingerprints for DNA and RNA detection, *Science*, *297*, 1536–1540

[17] Alivisatos, P. (2004).The use of nanocrystals in biological detection, *Nat. Biotechnol*, *22*, 47–52.

[18] Alivisatos, A. P.(1996). Semiconductor clusters, nanocrystals and quantum dots, *Science*, *271*, 933–937.

[19] Alivisatos, A. P., Gu, W. W. & Larabell, C. (2005). Quantum dots as cellular probes, *Ann. Rev. Biomed. Eng.*, *7*, 55–76.

[20] Pinaud, F., Michalet, X., Bentolila, L. A., Tsay, J. M., Doose, S., et al. Advances in fluorescence imaging with quantum dot bio-probes, *Biomaterials*, *27*, 1679–1687 .

[21] Michalet, X., Pinaud, F. F., Bentolila, L. A., Tsay, J. M., Doose, S., et al. Quantum dots for live cells, *in vivo* imaging and diagnostics, *Science*, *307*, 538–544.

[22] Gao, X. H., Yang, L. L., Petros, J. A., Marshal, F. F., Simons, J. W. & Nie, S. M. (2005). *In vivo* molecular and cellular imaging with quantum dots, *Curr. Opin. Biotechnol.*, *16*, 63–72.

[23] Smith, A. M., Gao, X. & Nie, S. (2004). Quantum dot nanocrystals for *in vivo* molecular and cellular imaging, *Photochem. Photobiol.*, *80*, 377–385.

[24] Chan, W. C. W., Maxwell, D. J., Gao, X.H., Bailey, R. E., Han, M. Y. & Nie, S. M. (2002). Luminescent quantum dots for multiplexed biological detection and imaging, *Curr. Opin. Biotechnol.*, *13*, 40–46.

[25] Action of clinically occult lymph-node metastases in prostate cancer, (2003). *N. Engl. J. Med.*, *348*, 2491–2499.

[26] Hood, J. D., Bednarski, M., Frausto, R., Guccione, S., Reisfeld, R. A., et al. (2002). Tumor regression by targeted gene delivery to the neovasculature, *Science*, 296, 2404–2407.

[27] Nie, S., Xing, Y., Kim, G. J. & Simons, J. W. (2007). Nanotechnology applications in cancer, *Ann. Rev. Biomed. Eng.*, *9*, 257−288.

[28] Cho, W. C. (2004). *Sci. J.*, *56*, 14−17.

[29] Cho, W.C. (2006). Research progress in SELDI-TOF MS and its clinical applications. *Sheng Wu Gong Cheng Xue Bao*, *22*, 871−876.

[30] Cho, W. C. S. (2007). *Molecular Cancer*, *6*, 25.

[31] Yim, E. K. & Park, J. S. (2006). *Expert Rev. Proteomics*, *3*, 21−36.

[32] Singha, A.K., Pandeyb, A., Raib, R., Tewarib, M., Pandeya, H. P. & Shukla, H. S. (2008). Nanomaterials emerging tool in cancer diagnosis and treatment, *Digest J. Nanomater. Biostruct.*, *3*, 135–140.

[33] Bruchez, M., Jr., Moronne, M., Gin, P., Weiss, S. & Alivisatos, A. P. (1998). Semiconductor nanocrystals as fluorescent biological labels, *Science*, *281*, 2013–2016.

[34] Seydel, C. (2003). Quantum dots get wet. *Science*, *300*, 80–81.

[35] Alper, J. (2005). Shining a light on cancer research. *NCI Alliance for Nanotechnology in Cancer USA*.

[36] Frangioni, J. V. (2003). *In vivo* near-infrared fluorescence imaging, *Curr. Opin. Chem. Biol.*, *7*, 626–634.

[37] Parungo, C. P., Ohnishi, S., De Grand, A. M., et al. (2004). *In vivo* optical imaging of pleural space drainage to lymph nodes of prognostic significance, *Ann. Surg. Oncol.*, *11*, 1085–1092.

[38] Gao, X., Cui, Y., Levenson, R. M., Chung, L. W. & Nie, S. (2004). *In vivo* cancer targeting and imaging with semiconductor quantum dots, *Nat. Biotechnol.*, *22*, 69–976.

[39] Dubertret, B., Skourides, P., Norris, D. J., Noireaux, V., Brivanlou, A. H. & Libchaber, A. (2002). *In vivo* imaging of quantum dots encapsulated in phospholipid micelles, *Science*, *298*, 1759–1762.

[40] Parungo, C. P., Colson, Y. L., Kim, S. W., et al. (2005). Sentinel lymph node mapping of the pleural space, *Chest*, *127*, 1799–1804.

[41] Parungo, C. P., Ohnishi, S., Kim, S. W., et al. (2005). Intraoperative identification of esophageal sentinel lymph nodes with near infrared fluorescence imaging, *J. Thorac. Cardiovasc. Surg.*, *129*, 844–850.

[42] Stroh, M., Zimmer, J. P., Duda, D. G., et al. (2005). Quantum dots spectrally distinguish multiple species within the tumor milieu *in vivo*, *Nat. Med.*, 11, 678–682.

[43] Morgan, N. Y., English, S., Chen, W., et al. (2005). Real time *in vivo* non-invasive optical imaging using near-infrared fluorescent quantum dots, *Acad. Radiol.*, *12*, 313–323.

[44] Davis, S. S. (1990). Assessment of gastrointestinal transit and drug absorption, In: L.F. Prescott and W.S. Nimmo, (eds.), *Novel Drug Delivery and Its Therapeutic Application*, Chichester, UK: Wiley, 89−101.

[45] Chourasia, M. K. & Jain, S. K. (2003). Pharmaceutical approaches to colon targeted drug delivery systems, *J. Pharm. Sci.*, *6*, 33−66.

[46] Davaran, S., Hanaee, J. & Khosravi, A. (1999). Release of 5-aminosalicylic acid from acrylic type polymeric prodrugs designed for colon-specific drug delivery, *J. Controlled Rel.*, *58*, 279−287.

[47] Schacht, E., Gevaert, A. & Kenawy, E. R. (1996). Polymers for colon specific drug delivery, *J. Controlled Rel.*, *58*, 327−338.

[48] Chung, K.T., Stevens, S.E. & Cerniglia, C.E. (1992). The reduction of azo dyes by the intestinal microflora, *Crit. Rev. Microbiol.*, *18*, 175−190.

[49] Yamaoka, T., Makita, Y., Sasatani, H., Kim, S. I. & Kimura, Y. (2000). Linear type azo-containing polyurethane as drug-coating material for colon-specific delivery: Its properties degradation behavior and utilization for drug formulation, *J. Controlled Rel.*, *66*, 187−197.

[50] Shantha, K. L., Ravichandran, P. & Rao, K. P. (1995). Azo polymeric hydrogels for colon targeted drug delivery, *Biomaterials*, *16*, 1313−1318.

[51] Van den Mooter, G., Samyn, C. & Kinget, R. (1992). Azo polymers for colon-specific drug delivery, *Intl. J. Pharm.*, *87*, 37–46.

[52] Ghandehari, H., Kopeckova, P. & Kopecek, J. (1997). *In vitro* degradation of pH sensitive hydrogels containing aromatic azo bonds, *Biomaterials*, *18*, 861–872.

[53] Kakoulides, E. P., Smart, J. D. & Tsibouklis, J. (2000). Azo cross-linked poly (acrylic acid) for colonic delivery and adhesion specificity synthesis and characterization; *J. Controlled Rel.*, *52*, 291–300.

[54] Furda, I. (2000). Aminopolysaccharides–their potential as dietary fiber. In: I. Furda, (ed.), *Unconventional Sources of Dietary Fiber, Physiological and* In vitro *Functional Properties*, Washington D.C., American Chemical Society, *14*, 105–122.

[55] Ormrod, D. J., Holmes, C. C. & Miller, T. E. (2000). Dietary Chitosan inhibits hypercholesterolemia and atherogenesis in the apolipoprotein E-deficient mouse model of atherosclerosis, *Atherosclerosis*, *138*, 329–334.

[56] Yamamoto, A., Tozaki, H., Okada, N. & Fujita, T. (2000). Colon specific delivery of peptide drugs and anti-inflammatory drugs using chitosan capsules, *STP Pharma. Sci.*, *10*, 23–43.

[57] Agnihotri, S. A., Mallikarjuna, N. N. & Aminabhavi, T. M. (2004). Recent advances on chitosan-based micro- and nanoparticles in drug delivery, *J. Controlled Rel.*,*100*, 5–28.

[58] Calvo, P., Remunan-Lopez, C., Vila-Jato, J. L. & Alonso, M. L. (1997). Novel hydrophilic chitosan-polyethylene oxide nanoparticles as protein carriers, *J. Appl. Polym. Sci.*, *63*, 125–132.

[59] De, S. & Robinson, D. (2003). Polymer relationships during preparation of chitosan-alginate and poly-l-lysine-alginate nanospheres, *J. Controlled Rel.*, *89*, 101–112.

[60] Chen, Y., Mohanraj, V. & Parkin, J. (2003). Chitosan-dextran sulfate nanoparticles for delivery of an anti-angiogenesis peptide, *Intl. J. Pept. Res. Ther.*,*10*, 621–629.

[61] Ma, Z., Yeoh, H. H. & Lim, L. Y. (2002). Formulation pH modulates the interaction of insulin with chitosan nanoparticles, *J. Pharm. Sci.*, *91*, 1396–1404.

[62] Honghua, X. & Tiejing, L. (2005). The analysis of boundary functions of CMS reaction factors, *Nat. Sci.*, *3*(2), 15–19.

[63] Jabbari, E. & Nozari, S. (2000). Swelling behavior of acrylic acid hydrogels prepared by γ- radiation cross-linking of polyacrylic acid in aqueous solution, *Eur. Polym. J.*, *36*, 2685–2692.

[64] Kumar, M. N. (2000). A review of chitin and chitosan applications. *Reactive and Functional Polymers*, *46*, 1–15.

[65] Dornish, M., Kaplan, D. & Skaugrud, O. (2001). Standards and guidelines for biopolymers in tissue-engineered medical products: ASTM alginate and chitosan standard guides, *Ann. NY Acad. Sci.*, *944*, 388–395.

[66] Van de Vord, P. J., Matthew, H. W., DeSilva, S. P., Mayton, L., Wu, B. & Wooley, P.H. (2002). Evaluation of the biocompatibility of a chitosan scaffold in mice, *J. Biomed. Mater. Res.*, 59, 58–69.

[67] Athanasiou, K. A., Shah, A. R., Hernandez, R. J. & LeBaron, R. G. (2001). Basic science of particular cartilage repair, *Clin. Sports Med.*, *20*, 223–231.

[68] Madihally, S. V. & Matthew, H. W. T. (1999). Porous chitosan scaffolds for tissue engineering, *Biomaterials*, *20*, 1133–1142.

# INDEX

## B

## C

## D

## G



## H

## L



## M

## N



## O



## P

W



X



Y



Z